Moving Times

Julian Weber

Moving Times

Mobility of the future

 Springer

Julian Weber
Munich, Germany

ISBN 978-3-658-37732-8   ISBN 978-3-658-37733-5   (eBook)
https://doi.org/10.1007/978-3-658-37733-5

Planned by: Dr. D. Froehlich

This Springer imprint is published by the registered company Springer Fachmedien Wiesbaden GmbH, part of Springer Nature.
The registered company address is: Abraham-Lincoln-Str. 46, 65189 Wiesbaden, Germany

Preface

Why This Book?

project i became known as the think tank of the BMW Group primarily through the development of two highly innovative electric vehicles: the megacity vehicle BMW i3 and the plug-in hybrid sports car BMW i8.

Beyond vehicle development, however, project i was also tasked with clarifying fundamental mobility issues. In order to be able to offer coherent mobility solutions in the future, the future requirements for individual mobility were surveyed in pilot projects carried out worldwide, relevant framework conditions were noted, and the acceptance of the new solutions derived from this was analyzed, for example: Where do I charge if I have no fixed charging facilities either at work or at home? How suitable are electric vehicles for car sharing? Can I sensibly reuse used batteries? How can I usefully include public transport and the electric kickboard in my boot in the route planning of my sat nav? In which major international city do which political and legal conditions for the registration or use of electric vehicles apply?

Being part of project i has been the most outstanding and personally formative experience of my professional life so far. Being part of the first line of this extraordinary think tank gave me the unique opportunity to deal intensively with all types of mobility stakeholders on an international level: Vehicle customers, old and new competitors, suppliers and service providers, new partners such as energy producers, journalists, investors, NGOs, representatives of municipalities and countries, and many more. I had the privilege of meeting many highly committed, creative, and innovative but, at the same time, critical people: car enthusiasts and critics, innovators and preservationists, believers and sceptics – and with every single one of these conversations, my personal picture of how people in different places around the world will get around in the future became more complete and coherent. To all of these people, my heartfelt thanks. Without them, this book would not have been possible, but it is also impossible for me to mention them all by name here.

My clear wish and claim as the author is to convey to you, the reader of this book, the emergence of this image of the future as understandable and comprehensible as possible – without, however, taking you out of the obligation to derive your own, individual picture from the information and opinions offered at the end. It is particularly important for me to emphasize at this point that the professional assessments and recommendations expressed by me in the context of this book reflect exclusively my personal opinion as a private person and in no way those of the companies and organizations mentioned therein, especially not those of the BMW Group or Clemson University.

Finally, I would like to point out that during the discussions and research for this book, I benefited greatly, both professionally and personally, from being able to exchange ideas with people of different origins, gender, skin color, and religion. In the twenty-first century, diversity and mutual respect should be a matter of course; I see both as extremely valuable enrichments in both my professional and private life. Against this background, it is important for me to note that although the generic masculine form is used throughout this book for the sake of readability, female and other gender identities are of course meant in the same way.

Munich, Germany   Julian Weber
February 2020

Contents

1

Introduction: Mobility in Transition

What Does It Depend on How We Will Get Around in the Future?

1.1 Mobility Needs and Supply

Mobility is life. To school or to work, to friends or to sports, on vacation or to discover the world – in any case, if you can't get around, you're limited in the truest sense of the word. Even though, despite many attempts, no formal right to mobility has ever been enshrined, there is a consensus in society that spatial mobility is a basic human need.

However, if we ask ourselves what mobility should look like today and, above all, in the future, the ideas are as different as the countries, cities, and people there themselves. Just as personal preferences in food, music, or home furnishings differ, so do individual mobility needs. Everyone has their own idea of when and where they want or need to go. In contrast to food or music, however, individual mobility needs cannot be catered individually; rather, mobility offers are geared to the overall needs of the population. This is just as true for public transport as it is for individual transport via the applicable rules and the necessary infrastructure. The analysis and forecast of individual and collective mobility needs, i.e., the question of who wants to go where and when and how many people there will be in total, is the content of Chap. 2.

J. Weber, *Moving Times*, https://doi.org/10.1007/978-3-658-37733-5_1

However diverse the needs may be, the basic supply-side elements of mobility are comparable all over the world: Cars on the one hand and public transport in local and long-distance traffic on the other represent the two supporting pillars. In addition, there are bicycles, scooters, tricycles, electric scooters, motorized rickshaws, and other vehicle alternatives that are often country- or culture-specific, as well as – spurred on by digitalization and the "use instead of own" trend that is spreading especially in international metropolises – a steadily increasing number of different mobility services in recent times.

1.1.1 Cars and the Automotive Industry

After critical years, the international automotive industry has now regained its economic footing to some extent, and the companies are quite successful overall. Nevertheless, in 2020, strategists, customers, and investors are more uncertain than ever about where the mobility business is headed. Evidence of this uncertainty can be seen in the unanimous announcements of change in the strategy papers published by car manufacturers. There is talk of "disruption" and "change", while "business as usual" is nowhere to be found. The realization is obviously spreading that digitization is not just a methodology for increasing the efficiency of business processes but is also fundamentally changing the expectations of customers and thus their buying behavior. Daimler's long-time CEO Dieter Zetsche, for example, became a consistent and credible advocate of the futility of strategic change at an early stage.

However, if you look deeper into the management structures of the automotive groups, you will see that the orientation toward and readiness for far-reaching change is still anything but universally anchored there. As is usual with disruptive processes, many of those in charge continue to cling emotionally to the established values, procedures, and artifacts that they consider to be causal for their personal success, despite a rational understanding of the need for change, and thus delay strategic further development and the associated maintenance of competitiveness, sometimes consciously, sometimes unconsciously.

Even when it comes to answering the question of what is actually required to implement this change, there is agreement across the houses of the industry, at least outwardly. The strategic priorities presented at the shareholders' meetings and published on the companies' websites are like two peas in a pod: Digital networking, autonomous driving, electric drive technology, and the provision of mobility services are identified by all relevant manufacturers as the four areas of expertise with which they want to equip themselves not only for survival but above all for success in the future of mobility.

It should therefore come as no surprise that the new players, whose business models are based on the breakthrough of these very technologies, are also proclaiming almost prayer-like that this is precisely where the journey will lead. Hardly anyone propagates the rapid and widespread spread of e-mobility as convincingly as a provider of innovative solutions for charging electric vehicles in an urban context; hardly anyone advocates the long-term superiority of autonomous driving over driver-controlled cars as convincingly and vehemently as the manufacturers of the sensors required for this. The strategic approach in these cases consists not only of their own conviction but also of a good portion of hope for the effect of the self-fulfilling prophecy.

Critical questioning is definitely in order here. To be sure, digitalization will dramatically change vehicle manufacturing via Industry 4.0 and car sales via direct access to vehicle and customer data – but will it have an equally dramatic effect on the cars themselves? That cars will one day actually be able to drive autonomously through cities and along motorways without a driver is extremely likely, if only because of the immense investment the automotive industry has made in this technology. However, are we equally sure that enough people will want to be taken to their destination in such a vehicle without a driver – quite apart from the question of whether it will be their own vehicle or not? Similarly, there is currently consensus across the industry that the vast majority of vehicles will be powered by electric motors in the future – but there is far less consensus on the question of what the ideal energy storage system for this will look like and who will pay for the necessary charging infrastructure.

All in all, the automobile manufacturers today have a clear picture of the technical solutions they will be working on in the coming years.

However, their ideas about how mobility as a whole, and thus ultimately also the demand for the vehicles and services they offer, will actually develop are far less concrete.

1.1.2   Public Mobility

In addition to individual car traffic, public transport is the second mainstay of mobility in both local and long-distance areas. Especially in the international conurbations, both regularly reach their limits during rush hour. You don't have to live in New York, Paris, or Tokyo to know that the situation on commuter trains every morning has nothing whatsoever to do with the advertising posters showing contented passengers relaxing on comfortable seats reading their newspaper on the way to the office. Nevertheless, in most large cities, it is precisely these train connections, which are referred to as "mass transit" in English, that ensure the connection of people from the suburbs to the core cities. In the cities themselves, an increasingly dense network of express trains, subways, trams, and buses takes people to their destinations.

Although not quite as dramatic as in the cities, the traffic situation on the roads is also becoming increasingly unbearable in long-distance traffic. Airplanes, trains, and now even long-distance buses are the alternatives to owning a car – although the latter are attractively priced, they ultimately also get stuck in the same traffic jams that you would get stuck in if you were driving your own car.

In contrast to local and long-distance transport in large cities, the transport problems outside the conurbations, in small towns and rural areas, are by far less pronounced, and the overall demand for mobility there is significantly lower. Small towns and villages have therefore always been rather neglected in terms of public transport provision, and many connections that still existed in the past have been dismantled over the years for cost reasons. As a result, there are hardly any real alternatives to owning a car in these areas, which limits the mobility of children, young people, senior citizens, and other groups of people who cannot, are not allowed to, or do not want to drive. Particularly in these areas, great hopes are therefore placed on new, private mobility providers.

Public mobility is therefore also undergoing major change, with the previously clear boundaries between public and private services becoming blurred. The forms, acceptance criteria, and future potential of public mobility are discussed in Sect. 5.4.

1.1.3 Mobility Services

The more the classic mobility and business model of "driving your own car" is questioned by the public, politics, and business, the higher the expectations of success for the new players in the mobility business become. In Germany, for example, companies such as the car sharing operators DriveNow and Car2Go, which have since merged to form SHARE NOW, the ride hailing provider Uber, and the long-distance bus company Flixbus – names that hardly anyone knew until a few years ago – have established themselves in a market that was dominated for decades by taxis, public buses, and trains and have added additional options to the delight of users. The public clearly perceives these services as an attractive and healthy addition to the existing mobility offer. However, the conditions under which people in urban, suburban, or rural areas will be willing to rely completely on these services and then dispense entirely with their own second or even first car will depend primarily on their availability and price-performance ratio.

While the taxi, bus, and rail companies, which have been overrun by the dynamics of this development, are still intensively dealing with it as former top dogs in some cases, mourning the good old days and demanding stronger regulation of the market by the public authorities to protect their own business from the suddenly existing competition, the new players are making targeted use of the opportunities offered by digitization to flexibly adapt to the wishes of customers, offer individual mobility at significantly lower prices, and conquer new market shares step by step by scaling their services.

In contrast to the car manufacturers, the change described above is leading to open enthusiasm among these new mobility providers – and not least among their investors. Their offerings and the business models behind them are the subject of Sects. 5.2 and 5.3. Whether the latter are also sustainable in the long term, however, remains to be seen: As with many digital services, the new mobility providers are often primarily geared toward rapid growth with short-term profit rather than sustainable success – which makes cooperation with them rather difficult for the municipalities concerned.

1.2   Strategic Framework

Whether in metropolitan areas, long-distance transport, or rural areas, neither the mobility needs existing in the population nor the mobility offers available to meet them are stable; both depend on a variety of framework conditions and, like these, are changing increasingly dynamically. How mobility continues to develop is primarily determined by a strategic framework provided by trends in the areas of demographics, infrastructure, technology, society, and legislation.

Site-Specific General Conditions
The topographical conditions within a coherent mobility area are generally stable, and the climate there also shows no changes that are directly relevant to mobility. The effects of global warming are of course relevant here, but they influence mobility primarily via social and regulatory developments; the increase in ambient temperature per se does not lead to any behavioral changes relevant to mobility. However, prevailing terrain and weather conditions favor or impede certain mobility solutions. In Barcelona or Rome, for example, motor scooters are an integral part of the mobility system due to the mild climate, whereas in Moscow they are comparatively rare. Also, while Copenhagen and Amsterdam have become the most bicycle-friendly cities in the world, not least because of their flat topography, a city like Stuttgart, which is riddled with inclines, has a much harder time convincing its citizens to ride a bike.

The spatial structure, i.e., the way in which the areas belonging to a mobility area are built on and used, also has a decisive influence on the need for mobility. Whether historically grown or deliberately planned, the spatial structure determines the distances that have to be covered between home and workplace or school, for example, and can be controlled to a certain extent through appropriate urban planning interventions such as the location of kindergartens, schools, shopping facilities, and doctors in residential areas.

At the same time, the population structure and population growth give rise to concrete but variable needs and thus requirements for the mobility systems on offer. For example, the congestion of commuter trains in many large cities is the result of the steady influx of people into the suburbs of the major cities.

The scope and condition of the infrastructure have a direct and massive influence on the suitability, feasibility, and thus also the acceptance of new mobility offers. If you can't find charging stations, you won't go electric. Where there is no stable supply of mobile Internet, no one relies on a mobility app. On the other hand, where well-developed roads and sufficient parking spaces are available, the demand and thus the acceptance of alternatives to the private car is close to zero.

The location-specific framework conditions of mobility areas are therefore manifold and are considered in detail in Chap. 3.

Technological Framework Conditions
Whether vehicles or services, new mobility offers arise from technological innovations. The acceptance of electric vehicles, for example, stands and falls with their range and thus with the energy density of the installed battery cells; the new mobility services that can be booked via app were only made possible by advances in digitalization. In the end, however, the decisive question is always whether innovative ideas such as autonomous driving or hydrogen drives can actually be implemented in series production, how well they meet the real needs of users, and how well they will ultimately be accepted by the market. With this in mind, the relevant technological trends in the automotive sector are dealt with in Chap. 4.

Social Framework Conditions
The technical developments are contrasted by the preferences and values existing within society but also reservations and aversions and their effects on mobility behavior. These are also changing and no longer as slowly as

they used to. Examples of relevant and global trends include the afore-mentioned "using instead of owning", the increase in online shopping compared to traditional shopping in stores, or the generally increasing importance of ecological and social sustainability. Chapter 6 deals with the question of which trends can be regarded as stable in the long term and what effects changed behavior patterns can have on mobility as a result.

Regulatory Framework

Last but not least, mobility worldwide is increasingly determined by laws and regulations at state or local level. While these can lead directly to new services, as in the case of the abolition of the rail monopoly on long-distance bus services in Germany or the permission to use e-scooters on public roads, restrictive measures such as driving bans for diesel vehicles in German cities or driving bans for scooters with internal combustion engines in China also have a lasting and in some cases immediate impact on existing mobility systems. Chapter 7 addresses the question of which trends make up the regulatory framework and what this will mean for mobility needs and offerings.

2

Mobility Needs

Who Wants to Go Where and When – and How Will This Change in the Future?

Being mobile is indisputably one of the basic human needs. Also, even if you can theoretically spend your whole life in the same place, mobility is the basis for education, employment, social contacts, and health, not to mention the opportunities it opens up for leisure activities and life fulfillment. This also includes the so-called *transport genesis*; the most important reasons why a person "sets off" are education, work, care, society, and leisure. A differentiated consideration of these "reasons for moving" is a prerequisite for a meaningful forecast of the direction in which mobility needs and modes of mobility will develop in the future.

2.1 Individual Mobility Needs

From a bottom-up perspective, the *individual mobility demand* refers to the sum of all journeys that a single person has to or would like to make on average in a given mobility area using a vehicle, i.e., excluding journeys on foot. Also, however different (i.e. "individual") these mobility needs may be from case to case, the reasons why someone wants to move to another place can be divided into a manageable number of categories.

© The Author(s), under exclusive license to Springer Fachmedien Wiesbaden GmbH, part of Springer Nature 2022
J. Weber, *Moving Times*, https://doi.org/10.1007/978-3-658-37733-5_2

The concrete manifestations of the needs then differ from individual to individual, although clear regional or even social patterns can often be discerned.

Likewise, although everyone has their own personal priorities (e.g., visits to restaurants are certainly not equally important to everyone), these categories can be sorted according to their importance. In the first place are the trips in connection with work and education (both the regular and the singular), then come the trips to cover daily needs and for medical care, and finally the trips that occur in the context of leisure activities. In this order, the demand categories are presented individually in the further course of this chapter. However, the focus of the analysis is not so much on the question of what share the respective category has in the total demand today but, whether and if so, on how the individual mobility needs of this category will change in the future – in order to then be able to estimate the change in the total demand.

Regular Attendance at Place of Work or Training
Routes to the workplace, school, vocational training, or university differ from all others in their regularity, for one thing. In most cases, this is the same way every day, there in the morning and back in the evening. Even if flexible working hours and mobile working are increasing in some areas, it is the morning and evening rush hours that push transport systems to their limits day after day worldwide, especially in conurbations. On the other hand, however, it is precisely these regular journeys to the place of work or training that are also without alternative: punctuality is a fundamental agreement in the majority of employment and training relationships, and failure to comply with it generally leads to consequences under employment law or disciplinary action. Time lost in traffic jams or overcrowded buses or trains is accepted more or less without complaint, because ultimately one has no choice. If the time required for these journeys becomes too long and cannot be reduced by choosing a different means of transport, the only alternative is to reduce the distance between the two – i.e., either to look for a new job or training place or to move house.

Wherever the main content of work or training is communication with people or interaction with computers or networked systems, physical presence at the place of work or training is not absolutely necessary, and it is possible in principle to reduce the distances travelled, if not completely. This applies in particular to classical office work but also, for example, to attending lectures or further training events. Technical solutions for *teleworking* and *telelearning* have been available for years, and their acceptance by employers and educational institutions is increasing all the time – it can therefore be assumed that the average individual mobility requirement for regular visits to the workplace or educational institution will decrease slightly in the future.

Cross-Location Work or Training
In addition to the regular journey to the place of work or training, there are extraordinary journeys to different locations. The extent depends on the individual professional task. A supermarket salesman, but also the head physician of an accident clinic, leaves his workplace comparatively rarely – apart from attending conferences or further training events – while consultants, purchasers, or construction project managers spend significantly more time on business trips than at their own desks.

If it is a matter of direct interaction with people or things on site – as is the case with care services or tradesmen – or if personal presence is actually required for other reasons, there is still no alternative. However, with increasing performance, security, and acceptance of video conferencing systems, for example, a large number of meetings will no longer require an on-site appointment in the future, and travel time and costs will be eliminated. A slight decrease in individual mobility requirements can therefore also be forecast here.

Professional Transport of Goods, Merchandise, and Persons
If goods, merchandise, or persons are to be transported from one place to another, it is not the driver's journey but the transport that is the focus of the trip, which is why this case is basically not a mobility need. However, since the vehicles used nevertheless contribute just as much to traffic and environmental problems, the transport of goods, merchandise, or persons is also included in this list of individual mobility needs.

Given the continued growth of online shopping, local shopping in downtown areas, and private ride services like Uber or Lyft, the need for such transportation trips will continue to grow. Autonomous vehicles will only be able to take over this task to a fraction and only for passenger transportation; rather, the number of drivers performing such transportation trips will increase.

Purchasing and Transport

In addition to the increasing use of delivery services, also for groceries and everyday necessities, more and more people in metropolitan areas are doing their shopping on foot whenever possible and thus in the neighborhood. The weekly shopping trip to the supermarket in their own car, filling up its trunk, is being replaced by a day's shopping in bags or backpacks at the small 24/7 market around the corner. The large supermarket chains have recognized this shop-local trend, and their city markets, which are geared to small-scale shopping, are sprouting up in the inner cities. Two circumstances are responsible for this trend: on the one hand, the desire for ecological and social sustainability and, on the other hand, the growing time and monetary expenditure associated with traditional shopping with a car today. Even if "shop local" is not possible everywhere, and one is not spared the trip to specialist shops that are not available in the neighborhood, the individual need for mobility for personal shopping will certainly decrease in the future.

Medical Care

The field of medical care is also undergoing significant change as a result of digitalization. Networking, suitable software, and sensor technology for measuring basic body values are increasingly enabling diagnostic scopes, visual or auditory examinations, and, of course, consultations to be carried out without the patient having to be on site at the practice. However, for many examinations and, of course, all types of treatment, a visit to the doctor's office remains necessary in any case. Also, many patients will continue to visit the nearest practice not only to save time but also to accept longer travel times in order to reach the specialist they consider to be the best. On the whole, therefore, the individual need for mobility in connection with medical care will probably decline only slightly.

Visiting Friends or Relatives

The mutual visit of friends and relatives is a central element of social and family life. While such visits are often feasible on foot in rural areas, the circle of friends and relatives is usually distributed much more widely, especially in urban environments, and getting there requires a private or public means of transport. That this will not change much in the future is just as obvious as the fact that these visits can hardly be replaced by Skype or Face Time; the individual need for mobility remains constant here.

Local Recreation

Leisure activities such as visits to restaurants, cinemas, theaters, sports clubs, or music lessons also focus on interaction with others, which is why it will naturally not be possible to replace the routes to these places with digital services. In contrast to journeys to work, education, or relatives, there is a much greater freedom of choice in terms of where to carry out these activities. What effort one wants to make to get there is something everyone can always decide for themselves. For example, if you don't want to spend an hour driving across town to go out for Indian food, you might rather look for a different but closer restaurant. The individual mobility requirements for such location-based leisure activities will thus remain at the same level.

Holidays

Mobility in connection with holidays differs significantly from other needs: People only go on holiday a few times a year, usually outside their own conurbation, with several people, a lot of luggage, and often the desire to be mobile at their destination. By and large, this model will continue to prevail, even if, for various reasons, there is a discernible trend in some regions toward spending holidays in one's own country. In contrast, however, holiday travel is set to increase significantly, especially in the emerging economies of Asia, led by China. The reasons for this are, on the one hand, the rapidly growing prosperity, which enables more and more people to go on holiday at all, and, on the other hand, a slow departure from a traditionally strict work ethic, which only allows holidays per se in rare cases. Regionally, therefore, a stronger increase can be expected in individual mobility needs for holiday travel.

Driving for the Sake of Driving
At the end of the priority scale is the journey as an end in itself, just for fun and without a specific destination. As a rule, such journeys are made by car, motorcycle, or bicycle; pleasure trips by bus or train are probably the exception rather than the rule. The ever-worsening traffic situation in both urban and suburban areas naturally dampens the pleasure of such journeys in the long term, while at the same time they are becoming increasingly difficult to reconcile with the growing demands for a sustainable life. In this category, a significant reduction in individual mobility needs must therefore be expected.

Even if digitalization and urban design mean that one or the other route will no longer be necessary in the future: The mobility needs of each individual will decrease only insignificantly – if at all.

2.2   Collective Mobility Needs

The individual mobility needs of all inhabitants and visitors of a mobility area together result in its *collective mobility needs*. Different mobility areas can be compared with regard to their collective mobility needs on the basis of three main criteria:

- The number of people with mobility needs.
- The spatial distribution of the points of departure and destination.
- The temporal distribution of requirements over the day, week, or year.

Whenever demand exceeds supply for at least one of these variables, existing mobility systems reach their limits. Well-known examples of this are the morning and evening rush hour, local traffic gridlock during major events, or the mega traffic jams on the trunk roads at the beginning of the school holidays.

The overall situation is becoming continuously more difficult: the influx into metropolitan areas is increasing rather than decreasing worldwide, and a large part of the housing created in this context to relieve the pressure is being built on the outskirts of cities, usually initially without the corresponding infrastructure. Reducing the collective demand for mobility, for example, through urban planning measures, and ensuring that the remaining demand is met effectively, while at the same time taking into account other political objectives such as safety and environmental protection, is one of the key challenges for states and municipalities.

2.2.1 Managing Collective Mobility Needs

As in many other areas, it is much simpler, more elegant, and more resource-efficient to reduce demand for mobility as far as possible before taking care of it. In purely theoretical terms, the simplest way here would be to counteract the increase in the collective demand for mobility in large cities by limiting the influx of people. However, for a variety of reasons, this would be neither politically desirable nor feasible worldwide.

The only remaining measure is to control the collective demand for mobility via the individual demand for mobility. Here, there are three basic levers available to states and municipalities:

* Promotion of Tele-X:
  Digital methods such as teleworking, telelearning, or telediagnostics avoid pathways but also require an appropriate technical, legal, and social framework. Only if a high-performance IT infrastructure is available, mature systems are offered, and the new forms of work and business are also socially accepted will they become widely accepted, and we can thus leverage their potential.

- Creation of a decentralized supply infrastructure:
  Through anticipatory or corrective structural planning, the local availability of shops, schools, kindergartens, or doctors' surgeries in residential areas, in the surrounding countryside, or even in rural areas can be promoted in a targeted manner so that it is no longer necessary to travel long distances to visit them.
- Time decoupling of mobility needs:
  By making working hours and shop opening times more flexible and desynchronizing school holidays and public holidays, the need for mobility cannot be reduced overall, but it can be distributed more evenly so that peaks in demand can be defused. It is obvious that the shifting of public holidays, for example, conflicts with other public interests.

Which of these measures is the right one varies from case to case. Criteria are the urgency (i.e., whether the desired effect is to be achieved in the short, medium, or long term), the cost-benefit ratio, and the available resources of the respective municipality or country. In addition, many of the measures, such as making shop opening hours and public holidays more flexible, conflict with other political objectives and must be carefully weighed up by those responsible within the framework of their respective mandates.

2.2.2   Meeting Collective Mobility Needs

Not only the control but also the coverage of the collective mobility demand is a political mandate of the federal states and municipalities, which they implement together with commissioned or independently acting private-sector companies. The public sector has three core tasks here:

- Operation of local and long-distance public transport:
  States and municipalities provide basic mobility services within a mobility area, for example, by operating bus and rail lines themselves or by commissioning other companies to operate them.

- Provision of transport infrastructure:
  In addition to roads, cycle paths, footpaths including bridges, and tunnels, this also includes waterways for passenger shipping, for example, as well as all facilities and systems for traffic control and for the energy supply of vehicles – both for public transport and for private transport.
- Providing the regulatory framework:
  Creation of a legal framework within which manufacturers of new innovative vehicle concepts and private mobility providers can operate in an economically viable manner.

While the first two points are classic municipal tasks, the third point represents a relatively new challenge. Who is allowed to drive e-scooters where? What requirements must the vehicle and driver meet in private ride-hailing? Legal certainty is a prerequisite for the development of innovations and thus a decisive enabler for innovative mobility solutions.

The most important prerequisite for the emergence of future-proof vehicles and mobility services must be created by the states and municipalities themselves – namely, a reliable regulatory framework in which innovative ideas and companies can flourish.

2.3 Shift in the Concept of "The Market"

The changes in individual and collective needs over the last few years have led to a shift in the concept of *market* in the field of mobility worldwide. In the automotive industry in particular, the term market traditionally refers to a country or economic area (such as DACH or ASEAN) served by a sales company. It is assumed that not only the regulatory requirements but also the customer requirements are the same within the market. These market-specific requirements are implemented through special country versions. Simple measures here are, for example, the labeling of operating elements in different languages or special color and material offerings. Particularly complex technical differentiations of passenger cars are, for example, right-hand drive vehicles, specific engine variants, or vehicles with an extended wheelbase.

However, demographic trends such as globalization and urbanization are accompanied – especially in metropolitan areas – by a cross-border convergence of lifestyles. As a result, customer requirements now differ less between individual countries than between different living spaces – even within a single country. For example, the mobility requirements of a family of four in New York City are already much closer to those of a family in Beijing than to those of a family in a small town in Nebraska or in western China.

Against this background, it will become increasingly important in the future to differentiate mobility products and services not so much by country but by type of mobility area – i.e., to launch a Mega City Vehicle for residents of international cities rather than a special China version or a ride-sharing service for rural areas rather than a US-specific offering.

3

Mobility Spaces and Mobility Systems

What Are the Framework Conditions for Mobility: How Will They Change in the Future?

In this book, *mobility areas* are defined as contiguous areas connected by transport infrastructure within which residents and visitors regularly move. These can be conurbations consisting of one or more cities and their direct catchment area but also small towns or rural regions. Long-distance links between cities also constitute mobility areas. As a rule, such mobility areas are under the control of one or more public administra-·tion bodies.

The totality of the means of transport used within such a mobility area to transport people (such as private cars or bicycles, public buses and trains, or also vehicles of private driving services), the associated transport infrastructure (such as roads, cycle paths, bridges or tunnels), the systems used to control vehicles (such as mobility service apps), and also the energy supply infrastructure required to operate the means of transport (such as filling stations or charging points) is referred to as a *mobility system.*

The mobility systems that exist today in the mobility areas have grown there over decades and in some cases centuries and are continuously developing, especially in the cities. Different mobility areas require different mobility systems, while similar mobility systems usually exist in similar mobility areas. For example, typical European cities with a historic old

© The Author(s), under exclusive license to Springer Fachmedien Wiesbaden GmbH, part of Springer Nature 2022
J. Weber, *Moving Times*, https://doi.org/10.1007/978-3-658-37733-5_3

town, core area, and commuter belt usually also have comparable mobility systems. The complexity of mobility systems increases exponentially from rural areas, where there is often no alternative to the private car, through small towns and suburbs to the centers of metropolises with underground trains, taxi fleets, bike-sharing stations, elevated roads, or tunnels.

The personal decision on how someone wants to move from one place to another depends on a complex web of individual and external factors. In order to be able to determine which mobility system is the best solution for a mobility area in this context, corresponding criteria for classification are presented in the following sections. For example, topological and climatic conditions make some mobility alternatives fundamentally more suitable than others. At the same time, however, the utilization and availability of the infrastructure, i.e., the road network, parking space, or local public transport, also determine the choice of individual mobility mode, whereby utilization depends on the ratio of supply to demand. Whether additional solutions such as private driving services or bike-sharing are offered depends in turn to a large extent on local legislation.

3.1 Geographic and Demographic Context

3.1.1 Topography and Climate

The topographic and climatic conditions of a mobility area enable, favor, or prevent certain mobility modes and thus represent fundamental and fixed framework conditions for the mobility systems possible in it.

In cities with extensive settlement structures, such as Los Angeles, walking is largely ruled out as a mobility alternative. Mountainous terrain, such as in San Francisco or Stuttgart, at least restricts the use of bicycles, and, like rivers, railway lines or expressways require high expenditures for the construction of bridges and tunnels.

Weather conditions such as low temperatures or precipitation not only spoil the fun of using open vehicles such as bicycles, mopeds, or scooters; they also dramatically change the mobility behavior of a large proportion of residents from one day to the next, as well as seasonally, and thus represent a significant uncertainty factor in the planning of mobility systems: For commuters, for example, safe bike lanes, parking, and additional trains must then be provided simultaneously, depending on whether they decide to bike, drive, or take the subway to work in the morning. Another effect of persistently low temperatures is the reduction of the range of electric vehicles – which noticeably limits their acceptance in cold regions.

While the contribution of vehicle emissions to climate change (discussed in detail in Sect. 6.1 in the context of sustainability) is already leading to a change in individual mobility behavior to a considerable extent, the actual magnitude of global warming does so only in very rare cases. Specifically, when people cycle to work instead of driving, they do so to reduce emissions – not because the greenhouse effect has made it noticeably warmer.

3.1.2 Spatial Structure

The *spatial structure of* a mobility area is another fundamental framework condition for mobility systems developing within it. The distribution of residential areas and workplaces, but also schools, medical facilities, shopping, and leisure facilities within the mobility area, is determined by the *average trip distance of* its inhabitants and their distribution via the available transport routes. In classic monocentric conurbations such as the Paris or London metropolitan areas, which have grown up around a central city, the majority of daily commuter traffic is concentrated in a star shape on the main roads and train connections to the city center, whereas in polycentric mobility areas such as the Ruhr region or Los Angeles, the traffic flow is distributed in a network across several subcenters.

Decentralization and the associated shortening of the average journey length can only be achieved very slowly by means of coordinated spatial structure planning by all the municipalities involved.

3.1.3 Population Size and Density

Mobility areas can be divided into *agglomerations, urban areas,* and *rural areas* according to their number of inhabitants and the distribution of population density within the mobility area. The formal criteria for this classification differ from country to country. In Germany, for example, populated areas with 500,000 or more inhabitants and a population density of over 1000 inhabitants per square kilometer are considered to be agglomerations.

Since the population density is usually calculated from the data of the registration offices, it basically refers to the place of residence and thus reflects the distribution for the case that all residents are at home – i.e., primarily at night. As a result, this figure neglects the contribution of visitors and transients, which is not exactly insignificant for mobility. Office and business districts that are teeming with people during the day (i.e., people who have come there in the morning and return home in the evening) thus formally have an extremely low population density, while the "dormitory towns" on the outskirts of conurbations are assigned a high value on paper here, although it is comparatively quiet there during the day.

The international trend of urbanization, i.e., the migration of large parts of the population of rural areas to the cities, is leading to an increase in population density in the core cities and, as a result, to a shortage and increase in the cost of living space there, with a simultaneous loss of quality of life. This in turn leads more and more people to move to the quieter and less expensive suburbs and, as commuters, to accept significantly longer journeys to work or education. If the public authorities do not take appropriate countermeasures, such as creating jobs and training places directly in the suburbs, the need for mobility will increase even more than the population itself.

The opposite examples, namely, areas with extremely high population density and at the same time relatively low mobility needs, are, for example, the ethnically developed districts of large American cities ("Chinatown"), which as a microcosm are hardly ever left by a considerable proportion of their population.

3.2   General Traffic Conditions

The individual decision on which means of transport to choose for the journey to work or other journeys is primarily based on the local, temporal, and financial availability of the possible alternatives. A train connection that does not run at the relevant time is no more an alternative than driving one's own car for someone who cannot afford a car of their own. Secondary criteria such as speed, comfort, safety, or status only become relevant when there are also alternatives to choose from.

The transport policy levers used by local authorities to control mobility-related decision-making in this context are the availability and costs of roads and paths, facilities and systems for traffic control and parking areas, as well as local public transport and private mobility services. The corresponding legislative options are presented and discussed in Sect. 6.4.

3.2.1   Roads, Cycle Paths, Footpaths

Whether the route in question is then travelled by car or scooter or by bicycle or on foot depends crucially on the available roads or cycle paths. Given the volume of traffic, the number, degree of development and quality of the roads, right-of-way regulations, and, if necessary, traffic control systems determine the possible vehicle throughput (and thus the frequency of traffic jams), comfort, and safety – not only for self-owned vehicles but also for public buses, taxis, or vehicles from fleets of mobility service providers that use these transport routes. In the case of topographical obstacles, tunnels and bridges enable direct and thus faster connections but require long planning and implementation times as well as high investment costs and, in the further course, also maintenance expenses and thus represent the bottleneck not only in terms of traffic but also in terms of time and finances in the case of necessary expansions.

3.2.2 Parking Areas

In addition, the availability of parking spaces is increasingly becoming a decisive criterion in the choice of mobility mode, particularly in city centers. Due to the shortage of space in city centers, the provision of more parking space in the form of multi-story or underground car parks is associated with high investments. This also makes private parking spaces scarce and expensive; in 2016, a parking space in London's Hyde Park Gardens sold for a record price of €450,000. At the same time, the shortage of parking spaces means that, in major cities, the proportion of time spent looking for a parking space is now as high as 70 percent of the total driving time. Anyone wishing to register a car in Tokyo must first prove that they have their own parking space for it – and hardly anyone even sets off there with their car without first making sure where they can park it at their destination.

3.2.3 Traffic Control

Today, traffic flow is controlled by means of fixed or controllable lanes, traffic lights, priority regulations, and speed limits. In this way, however, it is not only possible to optimize traffic within a mobility area; by controlling traffic in a targeted manner, individual mobility alternatives can also be specifically promoted (e.g., through separate lanes for taxis and public buses) or impaired (e.g., through long red phases on arterial roads), thus making alternatives such as public railways or the use of bicycles more attractive.

A quantum leap in traffic control would be the transition from decentralized to centralized vehicle guidance, as is already being considered in London and other metropolises. Here, drivers would no longer use their own navigation systems to decide for themselves which route to take to their destination and where to park; instead, once they have entered their destination, they would be given a binding route and a parking space by a central traffic management system. Similar to the material flow control in a logistics center, the central traffic management system in this way utilizes the available roads evenly and thus ensures an optimal traffic flow across the entire mobility area.

3.2.4   Mobility Services

In addition to owner-occupied vehicles, mobility services are the second core element of today's mobility systems. These include, on the one hand, sharing services in which an operator's vehicles such as cars, mopeds, bicycles, or e-scooters can be spontaneously driven without having to own them. In contrast to ride-sharing services, such sharing services are particularly interesting for users who like to drive themselves even without their own vehicle.

On the other hand, mobility services also include all possibilities of not driving oneself but being driven. These are generally state, municipal, and privately operated bus, rail, boat, and air services, as well as all types of private transport services. The more heterogeneous the mix of services here within a mobility space, the more flexible users are in choosing the appropriate mobility mode. For people who cannot, are not allowed to, or do not want to drive vehicles – such as children, senior citizens, or the disabled – such mobility services are also often the only way to be mobile.

Particularly in metropolitan areas with highly dense cores and the resulting difficulties in using their own vehicles, *local public transport (LPT)* takes over a large part of the mobility needs. Suburban trains bring people from the suburbs to the city center and back; subways connect the districts of the metropolises with each other, and trams and buses with a dense network of stops provide comprehensive connections. Mobility services in long-distance transport, i.e., connecting metropolitan areas with each other, include long-distance buses, long-distance trains, and flights. The decisive factors for the acceptance of these services are, on the one hand, the degree of spatial and temporal coverage of the route network and the resulting travel time for the respective case, as well as reliability, costs, safety, and comfort.

In this context, rail-bound vehicles have the advantage of not being stuck in traffic jams together with all the other vehicles on the roads and thus being able to transport large numbers of passengers quickly and reliably. At the same time, however, they are also associated with extremely high investments for the municipalities – especially in the case of underground construction projects such as the Stuttgart 21 underground station or the extension of underground railway lines – which is why, in addition to regional trains, buses, which operate flexibly on existing transport routes without major additional investments, are used almost exclusively in small towns and rural areas.

In the past, if you wanted to be driven spontaneously, without other passengers and on individual routes, the only and relatively expensive alternative to public transport was the taxi. Today, all over the world, a wide variety of private ride service providers such as Uber or Lyft are entering the market and complementing existing mobility systems. How open a municipality is to this, unilaterally removing protective regulations in the relevant laws and thus allowing supply competition in favor of its citizens, greatly influences the acceptance of alternatives to vehicle ownership and is thus a decisive element in the design of mobility systems.

The design of an optimally complementary and, from the user's point of view, seamlessly merging mix of public and private mobility services is one of the main levers for creating functioning and sustainable mobility systems.

4

Technological Trends
What Will the Vehicles of the Future Be Capable of?

Whether driving yourself or riding along, whether owning or sharing, the attractiveness of a mobility service always depends on the features and functions of the vehicles used. Anyone who wants to get a picture of the mobility of the future should therefore be able to get a picture of the vehicles that will then be on the roads. In order to be able to do this, one should in turn take a close look at the technological innovations that vehicle manufacturers are working on today. Also, there is a consensus across all manufacturers in the industry today about which technologies are relevant here: electrified drives, autonomous driving, connectivity, and mobility as a service are the core elements of virtually all strategy papers. BMW talks about the four ACES (autonomous, connected, electric, service) and Mercedes about CASE, and these four topics are also clearly visible in Volkswagen's 16 group initiatives.

In the following sections, these four *strategic trends in the automotive industry* are presented and discussed. In order to gain a better understanding of the mechanisms relevant to the future, we will also look back at where one or the other technical solution came from and why it was or was not able to become established. The focus is on passenger cars, but the statements made can also be applied to other vehicle types.

J. Weber, *Moving Times*, https://doi.org/10.1007/978-3-658-37733-5_4

4.1   Electromobility

On the one hand, it was the technical advances in the development of lithium-ion cells, while on the other hand, it was also the increasingly stringent emission limits for passenger cars that led automobile manufacturers to resume the development of electric drives from the beginning of this millennium. It very quickly became clear that this could not simply be a matter of replacing combustion engines with electric motors. A sustainable implementation of electromobility requires not only new drive concepts but also a profound change in vehicle concepts, energy concepts, and ultimately mobility concepts. Not only car manufacturers and their suppliers but also energy suppliers and grid operators are regularly examining the strategic relevance of electromobility for their business models. For the latter in particular, it represents an attractive new sales market on the one hand; on the other hand, however, it also requires the generation of renewable energies – because only with these can electric vehicles drive emission-free not only locally but also in the overall view.

4.1.1   The Return of the Electric Motor

At the beginning of the last century, at the dawn of the automotive age, the vast majority of motor vehicles were electrically powered. The biggest shortcoming of the combustion engine, which was already available at the time, was that it had to be started by means of a hand crank – a process that required strength and experience, and during which inexperienced drivers sometimes broke their hands. However, when internal combustion engines could be started by laymen with the invention of the electric starter in 1911, they suddenly became the predominant drive technology for road vehicles due to the greater range they could achieve and have remained so to this day.

In order to be able to offer a mixture of driving performance, space requirements, and manufacturing costs that is optimally matched to each vehicle type and size, manufacturers have developed a wide range of variants of internal combustion engine drive systems over time. Technically, these differ fundamentally in the following criteria:

- Number and arrangement of cylinders (in-line six-cylinder, V8 engine, etc.)
- Position and orientation of the engine in the vehicle (front-transverse, mid-engine, etc.)
- Turbocharging (turbocharger, etc.)
- Driven axles (front-wheel drive/rear-wheel drive/all-wheel drive)
- Transmission type (manual/automatic)
- Combustion process and fuel (petrol/diesel)

In addition to the size and performance of the engine, the combustion process selected is primarily decisive for vehicle emissions. Diesel engines, which are more expensive to produce, have up to 15 percent lower CO_2 emissions than gasoline engines but at the same time produce additional pollutants such as hydrocarbons (HC), carbon monoxide (CO), nitrogen oxides (NO_x), and particles such as soot or fine dust.

In order to be able to comply with the increasingly stringent global CO_2 emission limits in the long term, most car manufacturers therefore made the strategic decision toward the end of the 1980s to significantly increase the proportion of diesel vehicles in their fleets. To achieve this, however, diesel engines first had to be made "presentable": Whereas diesel engines had previously been rather gruff and designed primarily for torque and reliability, they now became smooth-running, powerful, and dynamic and the engine of choice especially for high-mileage vehicles such as company cars. The petrol engine remained the drive for smaller vehicles with low mileage on the one hand due to its lower price but also the drive for sporty models due to its performance characteristics.

In order to reduce the emissions of combustion engines even further, further technical measures were then implemented, such as the optimization of injection processes, heat management in the engine compartment, exhaust gas aftertreatment, cylinder deactivation, or the engine start-stop function. The high costs for the development of the new drive technologies as well as the investments in the production facilities required for their manufacture were borne at this time by the automobile manufacturers and their suppliers in the conviction that this would ensure that the statutory emission limits could be achieved in the long term.

The use of alternative fuels is another way to reduce emissions from combustion engines. For example, the combustion of natural gas or LPG produces significantly fewer pollutants than petrol or diesel, and the combustion of hydrogen even produces only pure water vapor as exhaust gas. Already at the EXPO 2000, a fleet of BMW 7 series limousines was on the roads, whose engine was running on hydrogen. Today, some vehicle manufacturers offer series models with engines designed for the use of natural gas or LPG. Despite this, however, the use of natural gas, LPG, or hydrogen as fuel for combustion engines that are thus lower-emission or zero-emission has never caught on; in Germany, for example, the market share of vehicles powered by natural gas was 0.11 percent in 2017, while that of vehicles powered by LPG was 0.13 percent. The reason for this is the limited range and, above all, the low coverage of mobility areas with corresponding filling stations.

In parallel with all these measures, possibilities of powering vehicles with electric motors instead of internal combustion engines were repeatedly investigated. However, a sensible technical implementation always failed due to the size and cost of an energy storage unit required for an acceptable range. The electric BMW 1602e, for example, which was used as the lead vehicle in the 1972 Olympic marathon in Munich, had a lead-acid battery weighing 350 kilograms, with which it achieved a range of just over the marathon distance of 42.2 kilometers.

However, at the latest when European Regulation 715/2007/EC came into force in July 2007, it was clear that the stricter Euro 5 and Euro 6 emission limits announced therein could not be met by further optimization of internal combustion engines alone. At the same time, the further development of lithium-ion technology meant that rechargeable batteries were available for the first time, the energy density and costs of which made it seem technically and economically feasible to offer electric vehicles with a range sufficient for urban traffic. The first mass-produced vehicles of this new generation of electric vehicles with lithium-ion batteries and a range above the magic limit of 100 miles or 160 kilometers were the Tesla Roadster in 2008, the Mitsubishi i-MiEV in 2009, and the NISSAN Leaf in 2010. The MINI E produced by BMW in 2009 was not sold, as was the BMW ActiveE in 2011, but was used exclusively in pilot projects in order to gain new insights regarding the technology used and user behavior. Electromobility and its potential as an alternative vehicle drive now received enormous public and political attention internationally, which also fuelled correspondingly high expectations. Since then, registration figures have been growing in all markets: In 2018, around 80 percent more electric vehicles were sold in China and the USA than in the previous year.

4.1.2   Types of Electric Vehicles

4.1.2.1   Design Criteria

Range has never been a problem with combustion vehicles. For city vehicles with rather low consumption, a relatively small amount of fuel carried is sufficient; typical business vehicles, which can also be driven several hundred kilometers, then have a larger tank. This can be refilled at any time and within a few minutes via a more or less dense network of filling stations. As far as range is concerned, a journey of just under 1000 kilometers from Frankfurt to Florence with one or two refuelling stops is easily possible, even in a small car.

The situation is different for electric vehicles: Batteries can store very little energy compared to gasoline or diesel, and the higher the maximum amount of energy that can be stored, the larger, heavier, and more expensive the battery. At the same time, charging the battery takes much longer than filling a fuel tank. The journey from Frankfurt to Florence in an electric vehicle would be completely unacceptable if it had to be interrupted by six charging stops, each lasting 2–3 hours. From this consideration, we can derive the three factors that should ideally determine the design of the drive system of an electric vehicle:

* Mobility needs: How far is the vehicle usually driven each day? Only in the city, between the city, and the surrounding countryside or also longer distances on the motorway? How often does this average range need to be exceeded? Planned or spontaneous? How critical is it if the planned arrival time is delayed?

- Charging options: Are there enough and sufficiently fast charging options available? Can the vehicle be fully charged at home at night? Is it possible to charge it very quickly if necessary?
- Availability of alternatives: If necessary, can mobility alternatives such as a second car or car sharing services be used if the range of the vehicle is not sufficient for a trip or the vehicle could not be charged sufficiently?

All three factors depend on both the individual situation of the vehicle user and the infrastructure at the respective vehicle location, which is why there is no general optimum for electric vehicle drives. Rather, several drive concepts have become established here in vehicle technology, i.e., different combinations of electric and combustion engines, energy storage systems, and transmissions.

4.1.2.2 Electrified Drive Concepts

The range as the critical characteristic of an electric vehicle par excellence depends primarily on the type and amount of energy stored on board. A corresponding tank is required to supply combustion engines with fuel, whereas there are three fundamentally different options for supplying electric motors with electrical energy, each of which can also be combined with the other:

- The storage of electrical energy in accumulators, i.e., batteries that have to be recharged at the end of the range.
- The generation of electrical energy directly on board, for example, through the flame- and emission-free "combustion" of energy-rich gases such as hydrogen in a fuel cell. The primary energy carrier, i.e., hydrogen, must then be stored and carried on board.
- The direct supply of electrical energy by conductive or inductive current collectors, i.e., via a supply line with sliding contacts (as in trams) or contactless induction coils (as in electric toothbrushes or smartphones).

In contrast to the classic vehicle with an *internal* combustion engine, an *Internal Combustion Engine Vehicle (ICEV)*, which is powered exclusively by an internal combustion engine running on petrol, diesel, or alternative fuels such as natural gas or hydrogen, all electric vehicles are driven additionally or exclusively by an electric traction motor. The components required for this are described in more detail in the next Sect. 4.1.3, and the drive concepts based on them are presented here:

Hybrid Electric Vehicle (HEV)

The so-called *micro-hybrid* and *mild-hybrid vehicles* are not so much drive systems in their own right as combustion vehicles with additional measures to increase efficiency. The necessary technical changes are relatively simple and inexpensive to integrate into existing vehicle concepts, but the contribution to reducing emissions is then also limited. In a hybrid, the combustion engine is switched off when the vehicle is at a standstill by means of an engine start/stop function and then restarted when the vehicle is moving on. In addition, the energy generated by the electric starter acting as a dynamo during braking (then called a *crankshaft starter generator*) is used to charge the starter battery – this is known as *recuperation* of braking energy. There is no electrical support for the vehicle drive in this case. However, one can only speak of a hybrid vehicle if it is driven by at least two different types of engine. A micro hybrid is therefore not a hybrid vehicle in the true sense of the word.

In the mild hybrid, on the other hand, the internal combustion engine is briefly assisted by an electric motor when the vehicle is being driven (so-called *boost*). Here, the alternator is used as the electric motor – in reverse mode of operation. For this purpose, the alternator must be designed significantly larger than usual and draws the required energy from a small auxiliary battery, usually not the starter battery. This battery is charged by brake energy recuperation in a similar way to the micro hybrid but, in this case, by the alternator rather than the starter. However, due to the relatively low output of the alternator and the equally low capacity of the battery, purely electric driving is not possible with a mild hybrid.

In contrast, the *full hybrid* uses an additional electric motor and an additional battery, which are dimensioned in such a way that at least parts of the journey can also be driven purely electrically. The harmonious interaction of the combustion engine and electric motor is made possible by the engine transmission control system. The first mass-produced and classic full hybrid is the Toyota Prius, which came onto the market at the end of 1997 and was initially able to drive up to 75 kilometers per hour with a 35-kilowatt electric motor and up to about 5 kilometers in pure electric mode.

The limited electric range of a full hybrid vehicle, which is also highly dependent on the individual driving profile, is offset by the disadvantages of the space requirement, the additional weight, and the considerable additional costs for the additionally required components. Moreover, because of the small contribution to emission reduction, in most countries the purchase of an hybrid electric vehicle (HEV) is not financially subsidized, in contrast to the purchase of a plug-in hybrid electric vehicle (PHEV) or battery electric vehicle (BEV). It is therefore to be expected that the HEV drive concept will no longer play a major role in the medium term.

Plug-In Hybrid Electric Vehicle (PHEV)

By using a significantly larger battery than an HEV and the option of charging it externally, the electric range of the PHEV is increased to around 50 kilometers. This means that, in most cases, daily journeys can be covered purely electrically, and the ranges required for longer journeys are made possible by the combustion engine, which is also present. In the Toyota Prius Plug-in Hybrid, for example, which went into series production in 2012, a gasoline engine with 73 kilowatts of power is combined with an electric motor with 60 kilowatt of power, resulting in a maximum drive power of 133 kilowatts. With a battery with an energy content of 4.4 kilowatt hours, the Prius Plug-in Hybrid then achieves an electric range of just under 50 kilometers.

When the combustion engine and electric motor are used synchronously, maximum power is available to the vehicle with very sporty drive characteristics. If the motors then also both drive different axles, an electronically controlled, road-coupled all-wheel drive can also be represented via hybridization. One example of such a sporty PHEV concept is the BMW i8, which has been in production since 2013 and in which an internal combustion engine with 170 kilowatts drives the rear axle and an electric motor with 96 kilowatts drives the front axle. Here, the maximum system output of 266 kilowatts enables acceleration from 0 to 100 kilometers per hour in 4.4 seconds, while the top speed in electric mode is 120 kilometers per hour and 250 kilometers per hour in combined mode. At the same time, the battery's capacity of 5.2 kilowatt hours allows a purely electric range of up to 55 kilometers, with a total range of 440 kilometers.

The consumption of electrical energy and fuel, and thus also the vehicle emissions of a PHEV, depends to a large extent on when the electric motor or combustion engine is used. In order to be able to control this according to individual requirements, different driving modes are available to the PHEV driver:

- In normal *charge depleting mode,* the vehicle basically drives electrically, but switches to the combustion engine as soon as the desired speed or acceleration requires it. In this driving mode, the battery is charged in between by brake energy recuperation but discharges over the total driving time. As soon as it is empty, the vehicle continues to drive using only the combustion engine.
- A special form of charge depleting mode is *e-mode*, in which the vehicle basically only drives electrically. The range, acceleration capacity, and top speed are then limited accordingly. The combustion engine is only switched on briefly when power is deliberately called up by kicking down (fully depressing the accelerator pedal).
- If the driver first has to cover a longer distance on the motorway, but only wants or is allowed to drive purely electrically at the destination, for example, due to local emission regulations, he or she selects *charge sustain mode.* In this mode, the vehicle maintains an adjustable, minimum *state of charge (SOC) of* the battery, for example, 80 percent, while driving with the combustion engine, thus maintaining the remaining electric range.

- In *recharge* or *sport mode*, the vehicle is always powered by the combustion engine, with the electric motor being switched on for boosting when power requirements increase. In phases of lower power demand, the combustion engine drives the electric motor, which then recharges the battery in generator mode.

As the available PHEV modes have a direct influence on vehicle emissions, international emissions legislation has been adapted accordingly. When measuring emissions in the new legally prescribed driving cycle, the *World Harmonized Light Vehicle Test Procedure (WLTP)*, a *Utility Factor (UF)*, is calculated for PHEVs, which describes the ratio between the electric and the combustion-engine driven portions of the journey. A UF of 100 percent corresponds to purely electric operation and a UF of 0 to purely combustion engine operation. The legislators are particularly critical of the recharge mode, because under certain circumstances it can lead to the PHEV no longer being charged at all but being operated as an HEV.

The additional expense required compared to an HEV, i.e., above all the larger electric motor, the larger battery, and the additionally required charger, is kept within narrow limits in view of the abovementioned significant improvements in electric driving. If part of the vehicle concept from the outset, PHEV drives can also be integrated into existing ICEV concepts with reasonable effort. The fundamental disadvantage of having to accommodate two drive units in the vehicle becomes less pronounced the more frequently both drives are used together. Where complete freedom from emissions is required, PHEVs will certainly be gradually replaced by BEVs, but for many areas of application and regions, PHEVs will remain the drive concept of choice in the long term due to their flexibility and independence.

Battery Electric Vehicle (BEV)

Finally, BEVs are purely electric vehicles in which no internal combustion engine is used. This means that, on the one hand, electric motors with a higher output are required, and on the other hand, above all, batteries with a significantly higher capacity. The advantages of the drive concept are primarily its simplicity (as a rule, e.g., a manual gearbox is no longer required) and the complete absence of local emissions. The major disadvantage is that the range is limited by the size of the battery, which is always a compromise between the achievable range on the one hand and weight, space requirements, and costs on the other. Low battery temperatures and powerful auxiliary consumers such as heating/air-conditioning units or even vehicle lighting have a negative effect on range. In correlation to the range, the availability of a dense network of fast charging options is decisive for the utility value of a BEV.

For a long time, the range of BEVs was limited to just a few models, but now there are a large number of different models on the market in the basic and premium segments with different engine outputs and ranges. With the Model S, Model X, and Model 3, Tesla is so far the only manufacturer that exclusively offers BEVs and offers its customers its own proprietary fast charging network for free use.

At the same time, the range of BEVs has developed in a completely different direction: While the first BEVs were certainly dynamic, but primarily rather rationally designed vehicles, the drive power possible via electric motors has also led to new, extremely sporty BEV concepts in recent years. In November 2016, for example, the Chinese company NextEV unveiled the Nio EP9, a super-sport BEV with a total output of 1 megawatt distributed across four electric motors of 250 kilowatts each, with which the vehicle can accelerate to 200 kilometers per hour in 7.1 seconds. The communication of such concepts alone is causing electric vehicles to lose their reputation among the public as rather boring common-sense solutions even faster, and even dynamically oriented drivers are beginning to take an interest in electrically powered vehicles.

BEVs are distinguished from all other drive types by three concept-related strengths: Complete local emission freedom, smooth running and dynamics. If individual driving profile, range and available charging infrastructure fit together as a framework, BEVs are certainly the drive concept of the future par excellence.

Extended Range Electric Vehicle (EREV)

For many people considering the purchase of a BEV, the fear of being stranded on the side of the road with an empty battery is a major reason for deciding against the purchase in the end. As a remedy for this *range anxiety*, electric vehicles are offered with a *range extender*, a type of emergency generator built into the vehicle. This consists of a compact combustion engine with an attached generator that can be used to recharge the vehicle battery or maintain its charge status as needed. In contrast to the PHEV, the combustion engine of the EREV is not used directly to drive the vehicle but only to power the generator. This makes it possible to operate the combustion engine constantly at the optimum speed from a consumption and emissions point of view, regardless of the vehicle speed. Electric drives with range extenders are also referred to as *serial hybrids*.

As with the PHEV, the EREV also offers a choice between charge depleting and charge sustaining mode. In the latter mode, the range extender switches on as soon as the battery's state of charge falls below the predefined limit value and then keeps it constant.

Whereas PHEVs tend to be internal combustion vehicles which, by adding a comparatively small electric drive, enable electric and thus emission-free parts of the journey, EREVs stand for BEVs extended by a small internal combustion engine which, if the worst comes to the worst, prevent them from stalling. Regardless of the actual performance and emission data, EREVs are seen in the public perception as being significantly closer to the ideal of zero-emission propulsion than PHEVs, and this applies in particular to the question of eligibility for funding. Legislation in some countries, for example, has considered sealing the fuel tank of an EREV promoted as an electric vehicle so that the range extender is actually and demonstrably only used in an emergency.

As expected on the part of the automobile manufacturers, in recent years, with the increasing spread of electric vehicles and the associated experience knowledge regarding their range on the one hand and the increasing capacity of their batteries on the other, the range anxiety and thus also the demand for range extenders has fallen sharply. For example, while more than half of the vehicles ordered for the first-generation BMW i3, whose battery had a capacity of 22.0 kilowatt hours, still came with a range extender, the figure for the second generation, with 33.2 kilowatt hours of battery capacity, was already <10 percent. And for the third generation of the BMW i3, which now has 42.2 kilowatt hours of battery capacity, no range extender is offered at all. For cost reasons, range extenders are therefore not newly developed but are assembled from existing combustion engines and generators – which is why their overall efficiency is often surprisingly low.

Fuel Cell Electric Vehicles (FCEV)

A completely different approach to extending the range of an electric vehicle is to generate electrical energy on board the vehicle using a fuel cell. In this, hydrogen, or in rarer cases methane or methanol, reacts with atmospheric oxygen to generate electrical energy, which in turn is used to drive an electric traction motor. Such a cold combustion of hydrogen does not produce any pollutants but only pure water.

Since hydrogen has a significantly higher energy density than lithium-ion batteries, hydrogen and fuel cells stored on board can achieve significantly greater ranges than pure BEVs for a comparable system weight or size. Series-produced FCEVs such as the Toyota Mirai, the Honda FCX Clarity, or the Hyundai ix35 FCEV have a range of between 500 and 650 kilometers with an engine output of between 100 and 130 kilowatts.

However, the main problem in the development of FCEVs is and remains the transport and storage of hydrogen. Two issues are at the forefront here: firstly, the fire and explosion risk of hydrogen mixed with oxygen, especially in the event of an accident. The frequently cited feeling, stemming from the term "hydrogen" and the bomb-like shape of the tank, of having a kind of hydrogen bomb on board leads to irrational fears and in many cases to the rejection of FCEVs as a possible alternative. Secondly, the chemical properties of hydrogen make it difficult to store on board. Due to its small molecular size, hydrogen diffuses through a variety of materials, including steel, and is thus lost. At the same time, these materials are damaged by the resulting embrittlement. The hydrogen storage systems currently available in FCEVs are therefore relatively large and heavy, making them difficult to integrate into existing vehicle concepts – especially since an additional, albeit small, battery is needed to temporarily store the difference between the energy delivered by the fuel cell and the energy absorbed by the electric motor.

For on-board storage of hydrogen in FCEVs, two different technologies are basically used today:

- Pressurized hydrogen storage:
  Storage of gaseous hydrogen in tanks under high pressure, in the Toyota Mirai, for example, at 700 bar in a carbon fiber-reinforced plastic tank. The compression work required for storage is not performed in the vehicle but by the corresponding hydrogen dispenser.
- Liquid hydrogen storage:
  To accommodate larger quantities of hydrogen, for example, in buses or trucks, it is liquefied and stored at temperatures around -250 °C. On the one hand, this requires a great deal of effort to liquefy the hydrogen before refuelling and, on the other hand, to maintain the low temperature of the hydrogen in the vehicle tank.

In addition to the safety concerns already mentioned, the availability of hydrogen refuelling stations is the greatest obstacle to the spread of FCEVs. The effort required to build or expand a hydrogen infrastructure, from industrial production to distribution via pipelines or tankers to delivery to filling stations, is immense under the given framework conditions such as safety requirements, hydrogen diffusion, compression work, or low-temperature maintenance.

The future of FCEVs must therefore be assessed in a differentiated manner. The use of hydrogen-fuelled fuel cells, for example, is certainly a highly sensible alternative in the long term for buses in local and long-distance public transport that require large amounts of energy, have sufficient space for appropriate storage systems, and can be refuelled and serviced centrally at the operator's depot. However, as the performance and coverage of the electric charging infrastructure and the capacity of vehicle batteries increase at the same time, the range-to-cost ratio of FCEVs compared to BEVs or PHEVs is continuously deteriorating. It is therefore more than questionable whether fuel cells will be an alternative for privately used passenger cars in the long term. After several postponements, Daimler's GFC F-Cell fuel cell vehicle, for example, is not going on sale as originally announced but can only be rented by a limited group of people in the interests of controlled operation.

Because the capacity of batteries is increasing and the charging infrastructure is continuously being expanded, the advantages of FCEV concepts over BEVs or PHEVs are continuously dwindling. As suitable as fuel cells may be for buses and trucks: Whether they have a future in private passenger cars is more than questionable for me.

4.1.2.3 Architectures of Electric Vehicles

In addition to the question of which drive concept is now the right one for a given requirements profile, from the vehicle manufacturer's point of view, it is above all relevant how this drive concept can be optimally integrated into an overall vehicle concept. Both technical and economic aspects play a role here and – as is often the case – are in conflict with each other. Depending on whether the priority in system design is rather to increase technical performance or to reduce costs and economic risks, there are two alternatives for the joint design of drive and vehicle architecture:

Adaptive Design/Conversion

Anyone who wants to offer an electric vehicle with as little development effort as possible and thus as quickly as possible usually integrates the components required for the drive concept of choice into an existing vehicle concept. Although minor changes to the vehicle are unavoidable, it is primarily only the drive components that need to be adapted to the conditions in the vehicle.

Technically, such a *conversion* always represents a compromise. For example, the BMW ActiveE, which was not built for series production but only in small numbers for an early trial of e-mobility, was a BEV conversion. When the underlying base vehicle, the BMW 1 Series Coupé, was designed many years earlier, the ActiveE was far from being planned and thus not considered as a variant. The battery system of the BMW ActiveE therefore had to be divided into three parts and distributed throughout the vehicle: under the rear seat bench, where the fuel tank previously had its place; in the transmission tunnel, where the gearbox, drive shaft, and exhaust system had previously been housed; and under the front hatch, where the combustion engine had previously been and now where the engine mounts protruded functionlessly into the otherwise empty space. The electric motor, on the other hand, was relatively easy to integrate into the rear axle.

The one-off costs for changes to the overall vehicle design and additional safety tests are therefore kept within limits in the case of a conversion: In particular, the bodywork and chassis and, in the case of PHEVs, the basic engine can be adopted. The manufacturing costs per vehicle, on the other hand, rise sharply due to the necessary adaptation of components. A conversion as an electric vehicle concept, i.e., the realization of a BEV or PHEV on the basis of an existing combustion engine vehicle, is therefore often chosen when there is great uncertainty about the sales figures that can be achieved with this vehicle, and the associated risk of high investment is to be minimized. Even the only "BEV-only manufacturer" Tesla has acted in this sense and first launched a conversion in 2008, the purely electric Tesla Roadster developed on the basis of the Lotus Elise.

One aspect that should not be underestimated from a sales perspective is that a conversion does not represent a new vehicle in the customer's perception but merely a drive variant. For example, when the e-Golf was launched on the market, it was primarily perceived as a Golf, for which an electric motor was now available as a drive in addition to the familiar petrol and diesel engines. This in turn leads to cannibalization effects: If there had been no e-Golf, its buyers would probably not have bought a competitor vehicle, but a Golf with a different drive variant – such as a GTD, with a significantly higher model return. Secondly, the price cross-comparison across the different drive variants of a model hardly allows the higher manufacturing costs of the BEV variant resulting from the electric drive to be passed on to the purchase price accordingly; customers' willingness to pay more for a BEV is limited here.

New/Purpose Design

In contrast to the conversion, the vehicle is designed around existing, function-optimized drive components in the *purpose design*, whereby the desired vehicle characteristics such as dynamics, space, or consumption can be implemented in the best possible way. This approach is particularly suitable for BEVs, as the geometric conditions here differ significantly from those of an internal combustion vehicle: The electric motor, which is significantly smaller than the combustion engine; the elimination of the fuel tank, transmission, shafts, and exhaust system; and the large and heavy vehicle battery with temperature control and charger can be accommodated much better in the vehicle as part of such a new design.

Examples of early Purpose Design BEVs include the Nissan Leaf and the BMW i3 (advertising slogan "Born Electric"). Now that there is much more experience with electric vehicles and the economic risks are seen to be lower, more and more manufacturers are offering Purpose Design BEVs. Tesla plays a special role here, offering only BEVs – with the exception of the abovementioned Tesla Roadster from 2008 – which do not use any parts taken over from existing conventional vehicles. Other manufacturers, on the other hand, understandably also use components and solutions from existing kits in purpose-design concepts. The fact that this can sometimes lead to inconsistencies is shown by the example of the Audi e-tron: a coherent purpose-design BEV in which parts of an existing electrical system were used – with the effect that the driver is prompted to change the oil after about 30,000 kilometers by a warning light that is not really necessary for an electric motor.

In the case of PHEVs in particular, a vehicle architecture developed from the very beginning as a common platform for the optional creation of ICEV or PHEV brings technical and economic advantages; the subsequent integration of an electric powertrain into an existing ICEV concept is significantly more complex. Examples of such Purpose Design PHEVs include the PHEV variants currently offered in the Mercedes C-,

E-, and S-Class or the BMW 3, 5, and 7 Series. Where on the scale between "optimized for ICEV" and "optimized for BEV" the sweet spot for such a platform then lies must be determined by product management on the basis of the expected ratio of unit numbers of the two drive variants.

The technical advantages of purpose design are always offset by the high development costs required for the new design. At the same time, the risks of the long-term product planning associated with purpose design become smaller the more experience with electric vehicles is available. Since they are also always technically superior to conversions, which are subject to compromise, purpose design concepts based on specially developed platforms will displace conversion architectures in the foreseeable future for both BEVs and PHEVs. This is evidenced by the corresponding strategy shifts by Volkswagen and Toyota, among others. The VW ID 3, for example, which will be launched in 2020, is based on a pure BEV platform developed for all brands of the Volkswagen Group, in contrast to the e-Golf.

4.1.2.4 Electrified Two-Wheelers

Two trends have led to a significant increase in the supply of and demand for electrified two-wheelers in recent years: Firstly, the rejection of noise and pollutant emissions caused by motorcycles, scooters, and mopeds, which goes hand in hand with increasing environmental awareness, and, secondly, the possibility of compact electric motors integrated into bicycles to reduce the physical effort required in such a way that cycling becomes an effortless mobility alternative, even over longer distances and uphill gradients. There are clearly different preferences in the global markets here, but three main types of electrified two-wheelers can be distinguished: electric motorcycles and scooters, e-bikes and pedelecs, and small electric vehicle concepts such as e-scooters or e-boards.

Electric Scooters and Motorcycles

As a typical means of urban transport, scooters were the first two-wheelers to be widely available with electric drive. For the comparatively short distances in the city, they are a real alternative to the commonly used two-stroke engine without exhaust gas aftertreatment, which is characterized by high noise and pollutant emissions combined with high fuel costs.

While supply and demand are still rather limited today in the vehicle class between 4 and 11 kilowatts of engine power (this includes, e.g., the BMW C Evolution or the models from IO Scooter and Trinity), a huge market has already developed worldwide in the classic moped segment up to 4 kilowatts or 45 kilometers per hour top speed. In Chinese cities, where scooters still represent an important personal mobility and status level between bicycles and cars, scooters with combustion engines were banned by law as early as 2006 due to their high exhaust emissions, which then gave an enormous boost to the development and supply of low-priced electric scooters. Internationally, the range of scooters on offer today extends from the simple and inexpensive models of Chinese manufacturers with mostly very modern design on the one hand to the many times more expensive retro-style models manufactured in Germany or Italy such as the Kumpan 1953, the Unu, or the Vespa Elettrica on the other. They are all being used – especially in the cities – by more and more people as not only sustainable but often also faster alternatives to cars or public transport. Because they are emission-free and technically robust, and because a normal car licence is sufficient to use them, small electric scooters are also ideal for urban sharing services, as described in Sect. 5.2.4.

In contrast to motor scooters, motorcycles are not usually used for a specific purpose but for weekend or holiday trips for their own sake. Here, the focus is on the ride and not on reaching the destination. Accordingly, the range of "real" motorcycles with electric drive is still very limited today and does not come from the established manufacturers (who, even more than car manufacturers, perceive the combustion engine as an indispensable part of the emotional experience of their products) but rather from new brands such as Zero, Victory, or Lightning

Motorcycles. Show bikes like the Harley Davidson Live Wire, however, show that even brands that have traditionally defined themselves massively by the technology and, above all, the acoustics of their combustion engines cannot avoid the topic of electric drive. Also, the performance figures for electric motorcycles, much like passenger cars, are far beyond what internal combustion engine motorcycles can achieve: A Lightning LS-218, for example, has an output of 149 kilowatts, a torque of 227 newton meters, and a top speed of 350 kilometers per hour.

The vehicle concepts are similar to those of conventional two-wheelers, so the limited safety and the lack of protection against the weather remain as major obstacles to acceptance. The electric wheel hub motors investigated by some manufacturers offer potential in terms of active safety, as their use in the front wheel also enables all-wheel drive on two-wheelers.

Pedelecs and E-bikes

The possibility of being able to cover longer distances and uphill gradients on a bicycle without much effort, supported by a small electric motor, is highly attractive in both the leisure and mobility sectors and has caused the market for pedelecs to virtually explode in recent years. Like a normal bicycle, a pedelec does not need an insurance license plate, but the motor is only switched on as a support while pedaling and must not exceed a maximum power of 250 watts. The required batteries are usually so handy that they can simply be dismantled and taken along for charging. The pedelec concept is particularly attractive for bicycle variants for transporting loads and passengers, for example, taking children to kindergarten or delivering goods to shops in pedestrian zones. A special form of pedelec that is unique to date is BMW's X2City, an electrically assisted foldable pedal scooter for the city.

Unlike pedelecs, S-pedelecs and e-bikes are permanently electrically powered; you can pedal – but you don't need to. However, these vehicles are no longer regarded as bicycles but as mopeds and therefore require an insurance plate as well as a driving licence (and therefore a minimum age) and a helmet. The legally permissible maximum speed of 45 kilometers per hour can be achieved on an e-bike by the motor alone and on an

S-pedelec only by the motor and pedalling. The maximum motor power of an S-pedelec is 500 watts, that of an e-bike 4 kilowatts. However, a safety risk that should not be underestimated is that e-bikes combine the drive power and top speed of a moped with the handling and active safety of a bicycle – and are thus more likely to be perceived as a bicycle by other road users but also by the rider himself and thus underestimated.

E-Scooters and Other Small Electric Vehicles

The availability of small electric motors with high power density as well as fast positioning control systems for balancing has led to the offer of more and more new, innovative driving devices over the last few years: Segways, e-boards, electric uni- and bicycles, and e-scooters, e-kickboards, and e-skateboards blur the boundaries between means of transport, fun sport articles, and toys.

E-scooters, pedal scooters with electric drives that are handy and practical especially for getting around in inner cities, play a role here almost exclusively as elements of urban mobility systems. They are usually foldable, making them easy to transport and thus an ideal complement to cars or public transport – which is one of the reasons for the strong increase in e-scooter sharing systems in cities. In addition, the vehicles, some of which are quite expensive, can be taken into the home and not only stored there safely but also charged.

What stands in the way of the acceptance and wider dissemination of these alternative vehicle concepts in some countries, however, is the still uncertain basis in some cases for traffic law. In many places, national and local legislators have been overtaken by technical developments and their rapid spread in the market and have since been busy clarifying whether these vehicles may be driven at all and, if so, on which public roads and under which conditions, such as minimum age or helmet obligation. The legal situation is thus conceivably inconsistent: in many US and European cities, e-scooters are already legally on the road today. In Germany, the Ordinance on Very Small Electric Vehicles has allowed e-scooters to be driven at a maximum speed of up to 20 kilometers per hour since 2019, analogous to the use of bicycles – i.e., on the road or on the cycle path.

The prerequisite is an insurance license plate and a minimum age of 14 years. However, vehicle concepts without handlebars, such as e-boards, may still not be operated in public spaces. In many major Chinese cities, the operation of e-scooters, which were already widespread at the time, was banned on public roads in 2016 for reasons of road safety.

4.1.3   Core Components of the Electric Drive System

From a technical perspective, an electric drive is primarily far less complex than an internal combustion engine drive. To drive a vehicle electrically, the required components known in the industry as the *Big Three* are the electric motor, which transmits the propulsive force to the road via its torque; the battery system, which stores the energy required for this on board; and the power electronics, which regulate the flow of current and the interaction with the on-board electrics and electronics. In addition, a charger is needed to charge the battery system.

Some classic automotive suppliers such as Bosch, Continental, or Siemens have strategically focused on the development and production of compatible components for electric drives. Future potential here lies in further integration of the individual components, in the aforementioned use of wheel hub motors (which eliminates the need for shafts, transmissions, and differentials), and in the development of low-speed electric motors, the use of which eliminates the need for a transmission.

4.1.3.1   Electric Motors

Compared to combustion engines, electric motors are characterized by their much simpler design, significantly lower power-to-weight ratio and significantly lower cooling requirements, which is why they are smaller, lighter, and also cheaper with comparable performance. In the case of electric vehicles, this basically offers the possibility of using several small motors integrated into the wheels, the so-called wheel hub motors. In most of the electric vehicles offered today, however – as in conventional vehicles – a central traction motor positioned at the front or rear of the vehicle is used, whose drive torque is distributed to the drive wheels via shafts and intermediate gears. However, especially when dynamics are no longer the top priority in vehicle design, wheel hub motors have the potential to be the drive technology of the future.

Recuperation

A specific feature of electric drives is the aforementioned ability to recuperate braking energy in overrun mode: when the driver takes his foot off the accelerator pedal, the electric motor switches from drive mode to generator mode, thereby decelerating the vehicle without braking. The resulting electrical energy is stored in the vehicle battery, which then recovers some of the energy previously used to accelerate the vehicle. The lossy use of the vehicle brakes is only necessary if the vehicle has to be decelerated particularly strongly, for example, in the case of emergency braking. Brake energy recuperation thus significantly improves the already good energy balance of the electric motor. For example, if you accelerate hard in city traffic after the traffic light has turned green and have to stop again immediately at the next red light, you get back a large part of the energy used for acceleration, in contrast to a combustion engine vehicle.

Types of Electric Motors

In electric vehicles today, three different types of electric motors are used as traction motors:

- In *permanently excited synchronous machines*, such as those used by BMW and VW, the magnetic field of the motor is generated by permanent magnets. These motors are characterized by their high efficiency and their low power-to-weight ratio of approximately 0.7 kilograms per kilowatt, making them light and compact. Speed control, speed reversal for reverse driving, and recuperation are easy to implement. A not insignificant strategic risk factor here is the fact that the so-called rare earths such as neodymium are required to manufacture the permanent magnets, the extraction of which can be critical from an ecological point of view, and the availability and price of which cannot be guaranteed in the long term today.

- *Separately excited synchronous machines*, such as those used by Renault, use electromagnets rather than permanent magnets to build up the magnetic field in the rotor. The sliding contacts required for their connection are subject to mechanical wear and must be replaced as part of regular maintenance.
- *Asynchronous machines* such as those used at Tesla are larger and heavier than the permanently excited synchronous machines, but they are also cheaper and more robust. They also enable the output of a multiple of their nominal power for short periods of time and thus extreme short-term acceleration values. On the other hand, a comparatively high degree of control is required. Since the magnetic field in the rotor is generated by induction, neither permanent magnets nor sliding contacts are required here.

4.1.3.2   Energy Storage Devices

As advantageous as the properties of the electric motor may be in converting energy into motive power compared to the internal combustion engine, the situation is exactly the opposite when it comes to supplying energy to the engine: While gasoline or diesel fuel can be provided in sufficient quantities in the vehicle tank with relative ease, the provision of electrical energy is still the main technical problem in the design of electric vehicles.

Battery System

In principle, storing the required electrical energy in a *battery system* (also known as *high-voltage storage* or HVS for short, especially in the case of BEVs) is the technically simplest way to supply the engine of an electric vehicle with electricity while driving. The challenge in practical implementation, however, is to resolve the conflict between the space requirements, weight, charging time, and price of the battery on the one hand and its capacity on the other. The vehicle range, which is limited by the battery capacity, and the effort required to charge the battery are still the biggest obstacles to the acceptance of electric vehicles.

This problem becomes a disadvantage for electric vehicles, especially in comparison to the properties of conventional vehicles with combustion engines. The relevant variable of an energy storage device here is its energy density – i.e., the maximum amount of energy it can store and thus also release again, divided by its mass. While the energy density of gasoline is 12.8 kilowatt hours per kilogram, the lithium-ion cells usually used in the battery systems of electric vehicles have an energy density of only 0.2 kilowatt hours per kilogram. This means that the battery system of such a vehicle weighs more than a 100 times as much as a full gasoline tank designed for the same range. Conversely, with the same weight of energy storage, the range of a vehicle with an internal combustion engine is more than a 100 times that of an electric vehicle.

As the range increases, the weight of a battery storage system becomes extremely high; in the case of the 85-kilowatt Tesla S, for example, it weighs a whopping 540 kilograms for a nominal range of 430 kilometers in the American EPA driving cycle. In contrast to the fuel tank, which becomes lighter as its capacity decreases, the weight of a battery system is independent of the state of charge; an empty battery weighs just as much as a full one. Another disadvantage is that the possibility of fitting battery systems into available vehicle cavities is severely limited compared to the free shaping of a fuel tank. Each battery is therefore ultimately a technical compromise that restricts essential vehicle characteristics.

Technically, a vehicle battery system consists of several interconnected battery cells, a device for cell temperature control (cooling or heating), the control electronics, and a housing that holds all these components together and protects them. Battery systems must be optimally integrated into the corresponding electric vehicle in terms of function and – especially in the case of conversions – geometry. They are therefore usually developed and manufactured by the vehicle manufacturer, using purchased, standardized battery cells.

Battery Cells

The actual energy storage takes place in the battery cells. State of the art here are lithium-ion cells – cells in which the cathode, the negative pole of the cell, consists of a material mix of lithium and other active substances such as cobalt, iron phosphate, nickel, or manganese. The anode, the positive pole of the cell, is usually made of graphite. This cell chemistry, with its range, safety, power, size, and weight, determines the relevant technical properties but also the costs of the cell – and thus also of the higher-level battery system. Lithium nickel manganese cobalt (NMC) has established itself as the cathode material for use in electric vehicles due to its property profile, while lithium iron phosphate (LFP) is another widespread, cost-effective alternative.

Over their useful life, lithium-ion batteries are subject to chemical wear, known as *degradation,* which continuously and irreversibly deteriorates key cell properties such as their capacity, voltage level, and self-discharge. Since the range of the vehicle is reduced along with the battery capacity, degradation directly affects its utility value. Measures such as optimized control of charging and discharging current, including even load distribution across all cells of a battery (so-called *cell balancing*), or maintaining an optimum temperature range of the cell during charging and discharging by actively heating or cooling the cell can slow down this effect but not stop it. Once the remaining capacity falls below a certain level (typically 80 percent), the entire battery or – if constructively possible, as in the BMW i3 – the affected battery modules are replaced. Replaced batteries and modules can then be reused as part of the "second life" in stationary operation, for example, as buffer storage or for frequency stabilization in the power grid, where energy density does not play a major role. This not only avoids recycling costs for used high-voltage batteries but may even generate a limited amount of revenue through their sale. The companies Audi and Umicore, for example, have demonstrated in a joint research project on the recycling of electric vehicle batteries that over 90 percent of the valuable materials such as cobalt and nickel can be recovered from the cells.

In addition to the cell chemistry, the shape and size of the cell are also relevant. The most common cells today are cylindrical round cells of the 18650 type, which are also used in laptops or mobile power tools. The type designation 18650 describes the dimensions of the cell: 18 millimeters in diameter and 65 millimeters in length. Such cells are used in Tesla vehicles, among others. In addition, there are prismatic automotive cells specially developed for automotive engineering – for example, in the BMW i3 – as well as cushion-shaped pouch cells, such as those used in the Chevrolet Volt. From the point of view of vehicle development, in addition to price and energy density, the flexibility that the cell shape allows in the geometric design of the battery system is particularly relevant. For example, a particularly flat battery integrated into the vehicle floor – such as is required for sporty, low-profile vehicles – can hardly be realized with the relatively large automotive cells. The shape and size of the battery cells therefore restrict the possible vehicle proportions and thus also the scope for vehicle design and thus have a high, often underestimated influence on the acceptance and economic success of the vehicle.

Lithium-ion cells have established themselves today as the solution for traction batteries in electric vehicles. The necessary raw materials are available for the foreseeable future, and the cells are fully recyclable. Due to the high one-off costs for research, development, and production facilities as well as the required number of units, an economic production of battery cells is hardly possible even for large automobile manufacturers. Instead, the cells are supplied to car manufacturers as standard components by highly specialized manufacturers such as BYD, CATL, LG Chem, Panasonic, or Samsung SDI. While for a long time the majority of lithium-ion cells came from Japan and South Korea, it is now becoming apparent that in the future by far the largest share of battery cells produced worldwide will come from China. Western manufacturers of electric vehicles are countering the strategic risk of dependence on a cell industry that is wholly owned by Asia by entering into strategic partnerships with cell manufacturers, such as Tesla with Panasonic, or by building up their own expertise, as BMW has done.

At the same time, the development of new cell types with higher power density at falling costs is a prerequisite for a higher range and lower costs of electric vehicles – and thus crucial for their market acceptance. The development and production of battery cells is therefore the primary key technology of electromobility, which is why manufacturers are investing billions in research into new cell types. The focus of research today is on lithium-air cells, lithium-sulfur cells, and solid-state cells, all of which promise up to 20 times the cell capacity of current solutions. However, these technologies are not expected to be ready for series production before 2025. In addition to the higher energy density, this also includes proof of fire and explosion safety, service life, and insensitivity to temperature fluctuations and mechanical stresses.

Given promising concepts such as solid-state cells and the huge investments being made in cell research around the world today, I certainly expect to see some real quantum leaps in cell technology in the future.

4.1.3.3 Power Electronics

A power electronics component is used in the electric vehicle to control the electric motor with the battery voltage, to regulate brake energy recuperation and to connect the 12-volt electrical system to the vehicle battery. Surprising to many, this usually exceeds the motor in size, weight, and cost. In addition, a non-switchable one- or two-stage mechanical transmission is used to reduce the speed of the electric motor to wheel speed and increase torque in the same proportion. If wheel hub motors are not used, a differential distributes the torque of the motor to the wheels of the driven axles in the electric drive, analogous to the ICEV. The electric motor, power electronics, transmission, and differential are usually integrated compactly in a common housing.

4.1.4 Charging

4.1.4.1 Charging Infrastructure Requirements

At least in the mobility areas whose problems and future this book primarily deals with, no driver of an internal combustion vehicle today has to seriously worry about being left stranded on the road with an empty tank. On the contrary, in Germany and other countries, stalling for lack of fuel on motorways and highways is even considered a misdemeanor and can be punished with a fine. If you run low on fuel, there are still plenty of filling stations within easy reach, even with a fraction of the fuel reserve left, where the fuel supply can be completely refilled in just a few minutes at any time of the day or night. Also, even on longer journeys, you can confidently trust that you will always be able to find a place to refuel and thus continue your journey. However, this confidence in the availability of the fuel infrastructure had to grow over a long period of time. As recently as 30 years ago, even in Europe, a full spare canister was usually kept in the boot of a car alongside the first-aid kit and warning triangle.

The technical requirements for charging electric vehicles, on the other hand, differ significantly from those for refuelling combustion vehicles. The disadvantage of charging is that the amount of energy that can be transferred over a certain period of time (the so-called *charging power*) is many times lower than when refuelling. Due to the restrictions of the battery already described, the range of electric vehicles is also much shorter than that of combustion vehicles in most cases. A charging process therefore not only takes much longer than a comparable refuelling process, but it also has to be charged much more often than refuelled. The fundamental advantages, on the other hand, lie in the energy supply. Generally speaking, the generation and distribution of electrical energy is much easier to realize than that of petrol, diesel, or even hydrogen. Electrical energy can be generated not only in power plants but also relatively easily on a decentralized basis. No tankers or pipelines are needed for its distribution, but only technically much simpler power cables. The intense political discussions today on the subject of cable routes would certainly be different if the alternative to "cable runs through my village" were not "cable runs somewhere else" but "tanker trucks roll through my village" or "hydrogen pipelines run through my village". Power outlets, albeit with limited power, are available in every home or commercial building, making basic charging infrastructure already available nationwide. At the same time, the performance and price of charging stations are continuously improving, and even high-performance fast charging stations tend to be cheaper to purchase and operate than a petrol or diesel pump.

Public Charging

Years ago, both vehicle manufacturers and politicians (who are very interested in the rapid spread of electric vehicles) recognized that only a comprehensive and high-performance *public charging infrastructure* would enable electromobility without unacceptable restrictions and that its availability is therefore an essential prerequisite for the widespread acceptance of electric vehicles. The possibility of being able to charge electric vehicles at home or at work using simple charging cables was certainly an important enabler of the renaissance of electromobility described above. However, this option is only available to a minority of drivers, namely, those who have access to private charging facilities in their own garage or at work and then tend to cover short, typical urban distances – at most to the point where they can return to their personal charging point. In order to make electric vehicles attractive not only as second cars, not only in the city, and not only for users with their own charging facilities, additional charging facilities must be created that are accessible to all and that meet the following criteria:

- Known:
  A list of the nearest charging options or those along a planned route should be available in the electric vehicle or via a smartphone app.
- Close:
  Charging possibilities should be distributed so densely that no long distances have to be covered for charging.
- Fast:
  The charging process should be able to be completed in as short a time as possible. However, if the required duration does not matter, the cost and battery life should be the main consideration.
- Available:
  If possible, it should be possible to ensure in advance that the charging facility approached can also be used – i.e., that it is not located in an access-restricted area, is defective, or is occupied by another vehicle.
- Compatible:

The charging station should have the desired charging technology for the respective vehicle. Important in this context is the standardization of voltage levels, charging processes, and plug types. It should also be possible to achieve the desired charging power.

* Comfortable:
  The use of the charging station should be as simple as possible, both in terms of operation and in terms of handling cables and plugs, etc. A roof also contributes to charging comfort in bad weather.
* Sustainable:
  The electrical energy obtained from the charging station should be "green", i.e., generated as far as possible without emissions and risks for the population and the environment.
* Safe:
  The charging station should be easy and safe to operate even for inexperienced persons.

A distinction must be made here between the different requirements of owner-occupied vehicles and fleet vehicles. Owner-occupied vehicles are driven for an average of about 2 hours per day and remain parked for the remaining 22 hours. This is a time that can be ideally used for slow charging. A fleet vehicle, on the other hand, whether in the fleet or in car sharing, is used by many different drivers, and the idle time here is significantly shorter. The focus is then on the rapid re-availability of the vehicle and thus rapid charging.

Private Charging

However, there are still many of the users described above who have their own charging facility at home where they regularly charge their electric vehicle. Here, the vehicle is usually connected to the grid for many hours from arrival in the evening until departure in the morning, so charging can normally take place at relatively low power levels. A timer integrated in the vehicle or in the charger allows the use of cheap night-time electricity if necessary, as well as the avoidance of grid overload by charging during periods of low load.

Since for many private users the pursuit of sustainability is the primary reason for using electric vehicles, this aspect is also of great importance when it comes to the origin of the electrical energy used for charging. However, while sustainability can only be felt in the abstract when concluding a green power contract, it can be experienced directly when using private systems for generating electricity from the sun, wind, or water. Many users of electric vehicles therefore want to drive with their "own", self-produced renewable electricity and supplement their charging station, for example, with a photovoltaic system and intermediate storage. In this case, the focus is usually less on a profitability calculation than on the aforementioned striving for a sustainable lifestyle.

4.1.4.2   Billing

Not least because of the mineral oil tax, electric energy is significantly cheaper today than petrol or diesel fuel. Depending on the type of vehicle, a range of 100 kilometers requires roughly 10 kilowatt hours of electrical energy, which at an electricity price of 20 cents per kilowatt hour costs 2 euros. The amounts to be billed at a charging station are therefore in the low euro range, and the expense of billing individual charging processes via credit card, for example, as at pay-at-the-pump pumps, is disproportionate to this. The high cost of billing compared to the amount of revenue means that shops, banks, or restaurants often offer their customers free use of their charging stations, thus saving themselves the cost of billing and then treating the electricity costs incurred as customer loyalty expenses.

As a rule, however, the user of an electric vehicle should of course pay for the energy consumed at the charging station. This requires identification of the user and measurement of the amount of energy delivered. Flat rates for charging electric vehicles make this measurement unnecessary and are therefore often agreed within companies and organizations. However, they have not been able to establish themselves at public charging stations – not least because flat rates fundamentally contradict the idea of sustainability, which is part of the lifestyle of many drivers of electric vehicles.

Today, user identification for the purpose of accessing a charging station is usually done via a customer card, for which an account must have been opened with the operator and bank details must have been deposited beforehand. The customer is then billed monthly for the amount of energy charged. So that an electric vehicle owner does not have to open an account with every charging station operator, these operators have joined together to form networks that enable customers to roam in a similar way to mobile telephony, i.e., to use all the charging stations in the network with just one customer card. The operators settle the accounts among themselves; the customer does not notice anything. Examples of such charging networks are ChargeNOW or Intercharge in Europe and ChargePoint in the USA.

4.1.4.3 Intelligent Charging

Before a vehicle can be charged at a charging station, the driver or the vehicle must register and authenticate. The charging station connects to the operator's central server, where the data of all authorized users is stored, checks the authorization, and then releases the charging current if necessary. After the charging process is completed, the charging station reports the amount of energy delivered back to the server via the same interface, where the user's customer account is debited. Reservation functions or fault messages also run via the data connection between the charging station and the operator server.

Today, mobile phone networks/GSM, LAN/Ethernet, or WLAN are used as transmission technology. Since a WLAN or LAN connection is not available everywhere and at the same time the reliable functioning of mobile phone connections cannot be guaranteed, for example, in underground car parks or dead spots, data exchange via the existing cable to the power supply, the so-called *power line communication (PLC)*, is a cost-effective and promising solution. In this case, the connection to the Internet can be implemented at any point in the power line between the charging station and the energy generator, for example, in the house distribution.

In addition to ensuring the actual function, data security must also be guaranteed when exchanging data between the charging station and the operator server. In particular, it is important to prevent unauthorized users from logging on to a charging station using other people's identities and illegally obtaining energy at the station's expense, but unauthorized access to vehicle data and its manipulation with the aim of causing damage must also be ruled out.

In order for an electric vehicle or its owner to know where charging stations are located and whether they are compatible and currently available, there must be a connection between the vehicle or charging app and the operator server in addition to the data connection between the charging station and the operator server. Via this connection, the user can access the list of public charging options, display the relevant information on them, and – if this option is available – also reserve the desired charging station. In the case of private charging stations, this connection can also be used to specify when the charging process should begin – for example, as soon as favorable night-time electricity rates apply or when there is a surplus of renewable energy in the grid. Data is exchanged between the vehicle and the operator server via a mobile phone connection, which is established either via the vehicle's own SIM card or by pairing the vehicle with a smartphone.

The company Ubitricity is pursuing a different, interesting approach here. On the one hand, it offers municipalities, companies, and fleet operators relatively simple and thus inexpensive charging stations that have neither a built-in electricity meter nor any kind of data interface and are thus particularly suitable for widespread use in urban areas – for example, for installing charging stations in street lamps. Secondly, users of electric vehicles are offered energy contracts with which (and only with which) they can charge at these charging stations. This requires a special smart charging cable, which is provided by Ubitricity as part of the energy contract. When the charging cable is plugged into the charging station, the customer is automatically identified and authorized; the charging cable then measures the amount of energy delivered and reports this back directly to Ubitricity for billing via a mobile phone connection. This solution is thus also particularly suitable for drivers who want to charge different electric vehicles on one bill.

4.1.4.4 Financing of the Charging Infrastructure

In 1880, on her legendary journey from Mannheim to Pforzheim, Bertha Benz still had to buy the petrol for the onward journey at a pharmacy. Since then, the infrastructure that exists today for the distribution of fuels via filling stations has developed. However, the associated business model has changed significantly over the past decades: Whereas petrol stations used to earn their money mainly from the sale of fuel and booked additional income from travel supplies and vehicle accessories, they now generate the majority of their profits from the sale of food and beverages, not least because of the increase in taxes to almost 65 percent for petrol and almost 55 percent for diesel. The competitive advantage of the markets and cafés integrated into the petrol stations, such as AGIP Ciao or TOTAL Bon Jour, over normal supermarkets lies not only in the longer opening hours. The favorable cost situation (space and staff are needed to sell fuel anyway) also plays a role. Thirdly, the fact should not be underestimated that customers who come because they need to refuel also take the opportunity to take food, drinks, and sweets with them – even if they are offered there at comparatively high prices. Today, petrol stations are also cooperating with supermarket chains here, as in the case of the REWE-to-go markets in ARAL petrol stations. The filling station business model is a profitable one and is subject to healthy competition. The manufacturers of motor vehicles have therefore never had to worry about whether their customers will find enough filling stations at which they can fill up their vehicles.

Today, the charging infrastructure for electric vehicles is still far from this. Public charging stations are usually located at the side of the road or in public parking lots so that the operating costs can be financed exclusively through the sale of charging electricity. The low profit prospects of this business model, which is not very attractive, are the main reason why energy suppliers do not take the initiative to quickly build a nationwide charging infrastructure, as oil producers do. However, according to estimates by the Swiss Bank UBS, the nationwide development of charging infrastructure, which is necessary for the politically desired rapid ramp-up of electromobility, will cost up to 360 billion US dollars worldwide over the next 8 years. For this reason, more and more consortia of automobile manufacturers, energy suppliers, and states and municipalities are being found to jointly bear the costs of the development. In the long term, however, a charging infrastructure that is financially unsustainable without public support is certainly not a stable business model.

For the future, the operation of charging infrastructure must be integrated into viable business models analogous to the gas station model. If charging stations are strategically positioned near restaurants or shops, charging also has advantages over refueling: While the refueling customer literally stands at his vehicle for a few minutes with his hand on the nozzle, then pays, and releases the pump for the next customer, the driver of an electric vehicle has up to 20 minutes or more in which he can move away from the vehicle and charging station without worrying and devote himself to other offers during this time – a promising situation for the operators of the corresponding shops and restaurants. For example, McDonald's offers free use of charging stations at its motorway restaurants or ALDI at the car parks of its city branches. In both cases, it has been shown that customers who charge their vehicles tend to stay longer in the restaurant or shop and also consume or buy more.

The price of the charging process is determined not only by the amount of energy consumed but also by the time required and thus by the charging power. If you have time, you charge more cheaply than if every minute counts. The additional costs for the installation of fast charging stations can thus be passed on to users, at least in part.

4.1.4.5   Vehicle to Grid (V2G)

A sustainable approach to make the operation of charging stations more interesting from a business point of view is the interconnection of many electric vehicle batteries, which are connected to the energy grid at the same time via charging stations, to form a large, distributed energy storage system, known as *Vehicle to Grid (V2G)*. Such a dynamic storage network can contain several hundred or thousand electric vehicles. Which vehicles are actually part of the storage network is constantly changing: Vehicles that are connected to a charging station are added; vehicles whose charging process is terminated leave the network. The charging station operator then makes this dynamic V2G storage available to an energy supplier or grid operator for a fee while granting the owners of the charging electric vehicles a discount on the energy they purchase, for example.

In practice, this means that the participating drivers of electric vehicles grant the operator of the V2G charging stations the right to also discharge their vehicle's battery for a short time during the charging process within a certain state of charge range (e.g., between 45 and 55 percent). Of course, this is primarily useful if the vehicle is connected to the charging station for longer than the actual charging process would take – for example, if it can charge at home at night or at work during the day. The planned departure time at which the vehicle battery must be fully charged is communicated to the operator when it is plugged in, so that he knows how long he can use it as a buffer and at what point he must start the remaining charging process at the latest. Fears that the battery could be damaged by multiple charging and discharging during V2G have not been confirmed in V2G applications to date.

Energy suppliers and grid operators are showing great interest in such storage systems. Battery storage systems can be used to compensate for critical deviations between the power fed into the grid and the power drawn from it within seconds, to stabilize the voltage frequency or to cut power peaks in the grid that would lead to higher electricity tariffs for the user or to overloads for the grid operator. Due to the increasing feed-in of solar and wind energy, the demand for the first-mentioned short-term load balancing in particular is increasing rapidly for energy suppliers. This so-called *primary control power* (availability within a maximum of 30 seconds) is otherwise purchased from specialized service providers. The price for the provision of 1 megawatt of primary control power is currently between 10,000 and 15,000 euros per month, which illustrates the economic potential of V2G applications.

4.1.4.6   Charging Technology

Technical Framework Conditions

A charging station, technically correctly called *electric vehicle supply equipment (EVSE)*, has the basic function of converting the alternating voltage (AC) available in the local power grid to the voltage level of the vehicle battery, i.e., usually to a direct voltage (DC) of between 350 and 400 volts. In order to be able to meet the diverse, individual requirements such as speed, availability, or costs, two parallel, mutually complementary approaches are being pursued in the development and expansion of the charging infrastructure:

• The nationwide installation of low-cost, but power-limited, charging stations for charging without time pressure, such as at home at night, at work during the day, or while shopping. These charging stations use the existing alternating voltage network; the conversion to the direct voltage of the vehicle battery is carried out by a charger installed in the vehicle.
• The installation of high-performance charging stations at specific points, which enable the vehicle battery to be recharged quickly when required, in a manner comparable to refuelling an internal combustion vehicle. The conversion of the mains voltage into the required DC voltage does not take place in the vehicle but in the charging station, which is therefore technically much more complex and more expensive to purchase.

One of the technical challenges in the implementation is the different national low-voltage grids. In Europe, China, and India, for example, a *three-phase* alternating current grid (a so-called *three-phase grid*) with a voltage of 230 or 400 volts and a frequency of 50 hertz is available, whereby the three phases are evenly offset to each other. The AC grid in the USA and Canada, on the other hand, has a frequency of 60 hertz and is in phase; here, voltages of 120 or 240 volts can be tapped. In South America, Africa, and Asia, on the other hand, there are many different national grids with voltage levels between 100 and 127 volts at 50 or 60 hertz and between 200 and 240 volts at 50 hertz. In Japan, there are even two parallel networks with different frequencies and voltages.

In order to manage the variance on the network side on the one hand and the different connection solutions on the vehicle side on the other, automobile manufacturers and charging infrastructure operators have also standardized the *charging types* together with the plug connections. From the user's point of view, the following types can be distinguished:

AC Charging

The easiest – but also the slowest – way to charge an electric vehicle is to use a standard household socket. In this charging method, which is therefore also known as *granny charging,* the vehicle is connected to the socket via a *Mode 2 charging cable.* An *In-Cable Control Box (ICCB)* integrated into the cable signals the maximum permitted charging current to the vehicle's charger.

In residential and commercial buildings, several circuits (colloquially referred to as *phases*) are usually connected to an AC house connection, to which permanently installed loads such as ceiling lamps or electric cookers and also sockets are connected via distribution boxes. In each of these circuits, the current is limited by a fuse, usually to 16 amps. This also limits the maximum possible charging power, for example, to a maximum of 3.7 kilowatts at 230 volts mains voltage.

However, the building installation, especially in older buildings, is often not designed for the permanent use of this maximum current. Mains plugs and sockets and cables with a small cross-section or terminal connections in distribution boxes can heat up considerably under continuous load. To avoid the risk of fire, the charging current for charging an electric vehicle must therefore be reduced again to up to 10 amps, which then reduces the actually available charging power in the 230-volt network to 2.3 kilowatts and in the 120-volt network even to 1.2 kilowatts. To fully recharge a completely empty battery with a nominal capacity of 24 kilowatt hours, approximately 8 hours are required in the 230-volt grid and approximately 17 hours in the 110-volt grid. In this way, the first electric vehicles of the new generation with their relatively small battery capacities could still be fully charged overnight by single-phase AC charging, which promoted their acceptance among buyers who had their own parking space or garage with a socket. For vehicles with longer ranges, however, this simple method of charging is now only suitable for emergencies.

Higher charging capacities can be achieved with AC charging by using an EVSE, which, in contrast to connection to a socket, uses its own circuit that does not feed any other consumers, is designed for continuous operation at maximum amperage, and has been checked and approved by a qualified electrician for safety in continuous operation before commissioning. The vehicle is then connected to the charging station via a *Mode 3 charging cable,* which is either hard-wired to the charging station or connected to it via an *IEC Type 2 connector.* The Mode 3 charging cable not only transmits the charging current but also enables bidirectional data transmission between the charging station and the vehicle's charger.

In the private sector, charging stations are usually designed in the form of a *wallbox* mounted on the wall of the house or garage and connected to the house wiring. This enables the permitted charging current to be doubled to 32 amps even in single-phase operation, which increases the maximum charging power in the 230-volt grid to 7.4 kilowatts and in the 120-volt grid to 3.8 kilowatts. A further increase in charging power is then possible through the joint use of several phases of the respective AC grid: In the case of the three-phase grid with three phases of 230 volts each, which is common in Europe, the charging power can thus be increased to up to 22 kilowatts in the case of domestic installations and to up to 7.4 kilowatts in the case of the additive use of two 120-volt phases, which is common in the North American region.

Charging stations in the public sector, on the other hand, are usually installed as free-standing charging columns in car parks or at the roadside and can thus be connected directly to the electrical supply network. Significantly higher currents of up to 64 amperes are therefore possible here, which means that a maximum charging power of 43 kilowatts can then be achieved in the 3 × 230-volt three-phase network and up to 15 kilowatts in the double-phase 2 × 120-volt network.

Two international standards have been established for connecting the vehicle and the charging cable for AC charging:

* The *IEC 62196 Type 1* (so-called *YAZAKI plug*) for single-phase charging with up to 7.2 kilowatts of charging power. This plug type is common in North America and Asia and is used by Citroën, Chevrolet, Ford, Kia, Mitsubishi, Nissan, Opel, Peugeot and Toyota, among others.
* The *IEC 62196 Type 2* (so-called *MENNEKES plug*) for single-phase and three-phase charging with up to 43 kilowatts of charging power. This type is used as standard in Europe and many other markets and is used by Audi, BMW, BYD, Mercedes-Benz, Renault, Smart, Tesla, VW, and Volvo, among others. Type 2 plugs are required for communication between the vehicle and the charging station as part of Mode 3 charging.

Until a few years ago, Italian and French vehicle manufacturers still offered an *IEC 62196 Type 3* plug connection (so-called *EV Plug Alliance plug*) as a third alternative. However, this has now been completely replaced by type 2 connections.

DC Fast Charging

In order to be able to charge BEVs – and in exceptional cases PHEVs with a high battery capacity – so quickly that one can resume driving after a relatively short waiting time, significantly higher charging powers of over 50 kilowatts are required. These are made possible by DC charging, in which the charging station converts the supply voltage into a DC voltage with the voltage level of the vehicle battery, which can then be used to charge it directly.

In addition to voltage conversion, the charging station also controls the optimum course of charging current and charging voltage over the duration of the charging process. In order to prevent damage to the battery cells, the charging current is reduced in accordance with a predefined charging current curve as the *state of charge (SOC)* increases. The charging current curve always represents a compromise between charging the battery as quickly as possible but also as damage-free as possible. On the one hand, this means that the maximum charging power or rated power of the charging station is only used when the SOC is low and only for a few minutes; on the other hand, this means that only a relatively low charging power is available when the SOC is high, which is why charging times for DC fast charging are often specified for charging from 0 percent to 80 percent SOC. The remaining charge from 80 percent to 100 percent SOC then often takes longer than it took to charge from 0 percent to 80 percent SOC. How much time the charging process actually takes depends on the one hand on the total capacity and the current SOC of the vehicle battery and on the other hand on the rated power (maximum charging power) of the charging station. This may also be limited by the supply line of the charging station, especially if several DC charging stations are connected to it.

From the initially regional developments of the automobile manufacturers and charging infrastructure operators, four standards have been established worldwide to date, according to which electric vehicles can be charged with direct current:

- *CHAdeMO* was the first available standard for DC fast charging and was developed jointly by Japanese vehicle manufacturers and energy providers as a powerful supplement to the comparatively weak AC connection via the Type 1 plug. This means that CHAdeMO always requires a second charging coupling on the vehicle side. The associated plug is comparatively unwieldy due to its size and weight. CHAdeMO charging stations today generally have a maximum charging power of 50 kilowatts, in some cases up to 150 kilowatts. Charging stations with up to 400 kilowatts are planned for the future. The CHAdeMO standard allows bidirectional charging and is therefore suitable for V2G applications. CHAdeMO charging stations are available primarily in Japan but also in Europe and North America. Electric vehicles with CHAdeMO connection are offered by Citroën, Honda, Kia, Mitsubishi, and Nissan, among others, while some manufacturers such as BMW, Honda, and VW offer the CHAdeMO standard only in BEVs for the Japanese market. Tesla vehicles have an adapter that allows them to charge at CHAdeMO charging stations.

- The so-called *Tesla Superchargers* use the DC fast charging standard of the same name developed by Tesla for its electric vehicles offered from 2012. The necessary DC lines were integrated into the existing Type 2 plug for this purpose, deliberately abandoning the plug's conformity to the standard. Only Tesla vehicles can charge at the Superchargers installed worldwide by Tesla itself; the maximum charging power there is 135 kilowatts.

- The *Combined Charging System (CCS)* was developed as a new charging standard by European and North American vehicle manufacturers and has been available since 2011. Unlike CHAdeMO, CCS integrates the DC charging option with the existing AC charging connection IEC Type 1 (*variant CCS 1*) or IEC Type 2 (*variant CCS 2*) in a common plug pattern, so that only one charging connection is ever required on the vehicle. CCS charging stations currently have a maximum charging capacity of 50 kilowatts, but there are plans to build stations with 350 kilowatts and up to 1000 kilowatts. CCS charging stations are widespread in the variant CCS 1 mainly in North America and as CCS 2 in Europe. In Europe, every fast charging station must have a CCS 2 connection since 2017. A CCS connection is offered in the electric vehicles of European and American manufacturers such as Audi, BMW, Chevrolet, Daimler, Ford, Opel, or VW (mostly as a chargeable optional extra), but Asian manufacturers such as Honda or Hyundai now also offer their vehicles with this option.

- As part of the government standardization roadmap for electric vehicles in China, the DC fast charging standard *GB/T 20234* was developed there. GB/T 20234 is based on the IEC type 2 connector but is not compatible with other standards. DC fast charging connectors of both Chinese electric vehicles and electric vehicles imported into China must comply with this standard.

The further development and regulatory agreement of DC charging standards is not only aimed at the feasibility of technical advances in charging technology but above all at securing market advantages. Once a charging standard has been established in a market, it is primarily electric vehicles that can be charged with this standard that can be sold there. The agreement of the manufacturers involved in the development of CCS to offer CCS as the sole charging option in their vehicles is proof of this, as is the power of CHAdeMO in Japan and South Korea or the agreement of an additional DC charging standard by the Chinese authorities, which is certainly not absolutely necessary from a technical point of view. From the vehicle manufacturers' point of view, the clear goal is to agree on a standard internationally and thus be able to equip all electric vehicles and charging stations with the same solution. From the point of view of vehicle users (and charging infrastructure operators), on the other hand, it would be sufficient if one standard were to prevail within a coherent traffic area, i.e., Europe, North America, Japan, or China, so that every electric vehicle available on the market could be charged at every charging station.

Which will be the predominant charging standard in the future will be decided primarily in the European and North American markets, where both CHAdeMO and CCS are available today. It is highly unlikely that both standards will remain in parallel in the long term, given the associated costs for manufacturers and disadvantages for users. Also, although there are significantly more CHAdeMO than CCS charging stations installed there today, CCS seems to be establishing itself as the solution of the future, not least because of its technically superior communication options. It is questionable whether CHAdeMO would then remain the standard for the Japanese market. It is equally unlikely that Tesla will maintain its proprietary fast charging system if it is overtaken in terms of charging performance by both CHAdeMO and CSS in the foreseeable future. In the Chinese market, on the other hand, GB/T 20234 is already the undisputed number 1 with about 150,000 charging stations installed and regulatory protection.

A possible nationwide expansion of DC fast charging infrastructure would also open up the strategically highly interesting *DC Only* option for manufacturers and users of electric vehicles. This refers to the approach of relying exclusively on the availability of DC charging stations, thus dispensing entirely with the vehicle's charger and saving weight and costs accordingly. Just as an internal combustion vehicle can only be refueled at a gas station, a DC-only electric vehicle could only be charged at DC chargers.

Battery Changing Systems

Another way to quickly bring large amounts of electrical energy into the electric vehicle is to swap batteries that have run down for fully charged ones – analogous to the procedure for battery-powered tools or forklifts. In mathematical terms, this approach is particularly interesting for high-capacity batteries: the exchange of a 100-kilowatt-hour battery that has been drained to 10 percent SOC within 5 minutes corresponds to an average charging power of 1.1 megawatts, i.e., more than 20 times the value of today's fast-charging systems. It is therefore not surprising that the first battery swap systems were used in electric local transport buses in China. In the city buses converted to electric drive, several battery modules the size of a Euro pallet and with a capacity of around 100 kilo-watt hours are stored in the stowage space in the lower part of the bus and are easily replaced with the aid of forklift trucks after the fastening elements and electrical connections have been loosened.

Implementing this concept in a technically and economically viable way for electric passenger cars was the goal of the Israeli company Better Place, which already set up functioning changing stations in Denmark and Israel in 2009 and operated pilot fleets with Renault Zoe and Nissan Qashqai to prove the concept's suitability. At the exchange stations installed there, the power and coolant lines of the empty battery were first automatically disconnected from the vehicle via special couplings, and then the battery's retaining clips were opened, whereupon the battery could be removed downwards and brought to a charging station. At the same time, an identical, fully charged battery was taken from a storage facility, brought to the vehicle, attached, and electrically connected. The whole process took <5 minutes from the time the vehicle entered the changing station to the time it left.

There were two main reasons why the technically quite convincing concept was not ultimately successful and Better Place had to file for insolvency as early as 2013: Firstly, the electric vehicle manufacturers did not want to commit to a standard battery for understandable reasons; this would have massively restricted the design freedom of new electric vehicle concepts and the flexibility with regard to adopting possible innovative cell and battery technologies at this early stage of the ramping up of electric mobility. Secondly, against the backdrop of the high investment costs for the swap systems and batteries, Better Place was never able to prove the economic viability of the swap concept and thus also failed to attract sufficient investors to finance the planned global expansion. When it then became foreseeable that, with increasingly powerful charging systems, even high-capacity batteries could be charged ever faster, the end of Better Place was also sealed.

Although the idea of battery swapping itself is being taken up again and again, as is currently the case with start-ups such as Nio in China or Atmo in California, the disadvantages mentioned remain. Battery swapping systems could become established, for example, for regionally operated electric vehicle fleets whose business model requires extremely fast recharging but certainly not for individual mobility.

Inductive Charging

At the latest since smartphones and electric toothbrushes have shown how practical and safe inductive charging is, the desire has also grown among electric vehicle users to be able to charge their vehicles without having to handle dirty or wet cables and bulky plugs.

Commercially available inductive charging systems for electric vehicles consist of two flat coils, one of which is integrated into the ground or some kind of floor mat and connected to the charging station, and the other is recessed into the vehicle floor and connected to the on-board charger. With the inductive systems available today, charging powers of 3.8 kilowatts are common and up to 11 kilowatts are possible. The prerequisite for this is that both coils are correctly positioned in relation to each other, which means that the centers of the two coils should be <10 centimeters apart, if possible, when viewed from above, and the air gap between the two coils should be between 10 and 14 centimeters when viewed from the front or side. When positioning the vehicle correctly above the ground coil, the driver is assisted by displays and acoustic signals. At the same time, metallic objects or small animals must be prevented from entering the magnetic field between the coils.

Inductive charging is now primarily offered as a convenience feature for PHEVs in the private sector, for example, in the BMW 530e iPerformance. Retrofitting is also available for many other models. Arguments against the use of inductive charging in public areas are the limited charging power, the high investment costs for the weather- and vandalism-protected integration of the ground coil including supply lines into the road surface, and the still missing agreement on an industry standard supported by all manufacturers. In addition to the convenience aspect, inductive charging offers the advantage, especially for autonomous parking of electric vehicles at a charging station, that a plug connection no longer needs to be closed.

Dynamic Charging

In contrast to stationary charging at a charging station, with dynamic charging the electric vehicle is not charged while parked but while driving. In principle, this requires a charging infrastructure of some kind running along the road and therefore costly.

As with trams or trolleybuses, *conductive dynamic charging* establishes the electrical connection between the vehicle and the charging infrastructure via cables or power rails recessed in the roadway or positioned at a safe distance above the roadway and the vehicle, as well as sliding contacts or current collectors on the vehicle side. For obvious safety and practicality reasons, this solution can be ruled out for individually used passenger cars.

In contrast, *inductive dynamic charging* is based on induction coils embedded in the road surface, over which the vehicle moves with its coil integrated in the vehicle floor. Corresponding systems are being tested today by Qualcomm and Renault, among others. The idea is certainly elegant and convenient, but there are two decisive obstacles to its widespread implementation:

* The limited charging power:
  With the pilot applications presented today, a charging power of up to 3.6 kilowatts can be realized, with 20 kilowatts mentioned as a long-

term goal. However, this charging power is of course only available on those sections of the route that are connected to the charging infrastructure. In addition, the maximum charging power per vehicle is limited by the number of vehicles charging on a supported route section in relation to the maximum connected load of the charging infrastructure for this route section. This is contrasted by the energy consumption: In real operation, the consumption of an electric vehicle in urban traffic is typically in the order of about 15 kilowatt hours per 100 kilometers. At an average speed of 30 kilometers per hour, which is common in urban areas, this corresponds to an average power consumption of 4.5 kilowatts. The power delivered by the battery for a city trip is therefore significantly higher than the power absorbed during this trip, and this gap becomes even larger as the average speed increases. Inductive dynamic charging can therefore at best delay the discharge of the battery while driving, but it is not an option for charging.

- The required installation effort:
  The area-wide laying of coils in the pavement of new and existing roads, the electrical connection of the coils to the power grid, and the implementation of billing (charging vehicles must identify themselves at the charging infrastructure when they first make a connection and after each disconnection) require investments of a magnitude that, in view of the question discussed above as to who will bear the costs of expanding the public charging infrastructure, will certainly not be borne by any of the parties concerned.

The field of dynamic charging also includes quite exotic ideas such as the mobile chargers discussed in specialist circles: As with the aerial refuelling of aircraft, charging vehicles with large batteries are supposed to follow the vehicle to be charged, connect to it, and carry out a quick charge while driving. Here too, quite apart from the technical feasibility, the economic viability of the approach is highly questionable.

4.1.5 Summary: Advantages of Electromobility

Even though it can be read in all media that the future belongs to the electric drive, and one sees more and more BEVs and PHEVs especially on the streets of the metropolises, the vast majority of even newly registered vehicles still run on combustion engines. Also, it's not just nostalgia that's keeping new car buyers from turning to electric vehicles. The primary disadvantage of electrified vehicles compared to conventional internal combustion vehicles is still the combination of the limited range, the time required to recharge the battery, and the limited availability of public charging facilities. However, intensive work is being done on all three of these issues, and it can be assumed that the framework conditions will continue to improve in favor of e-mobility.

On the other hand, the key advantage – and main driver of the renaissance of electric vehicles toward the end of the nineties – is their local emission-free status. However, electric vehicles have several other advantages over vehicles with internal combustion engines which, in combination with the reduction of the abovementioned disadvantages, have the potential to help electromobility achieve a long-term breakthrough: the compact design, the plus in driving dynamics and driving comfort, and the lower operating costs.

4.1.5.1 Zero Local Emissions

The issue of emissions from electric vehicles has been the subject of much public and media debate. It is true: Only BEVs have no local emissions at all; for PHEVs, they depend heavily on the personal usage profile and charging behavior. A PHEV with a short electric range, which is predominantly operated on long-distance journeys and in sports mode with high boost percentages and is accordingly only sporadically charged, differs only slightly in terms of its emissions from a normal combustion vehicle. In contrast, a PHEV with an electric range of 50 kilometers, which is fully charged every morning and mainly used in urban traffic, can be permanently operated purely electrically and thus emission-free.

If only because of the claimed sustainability, it is certainly right to look at the system as a whole and also take into account the proportions arising in the generation and distribution of electricity when measuring and evaluating the emissions of electric vehicles. It makes a clear difference here, of course, whether the energy used for driving comes from an old coal-fired power station, a nuclear power station, or, say, a wind turbine. However, it is also a fact that when making an objective comparison with internal combustion vehicles, the entire system and thus the entire value chain from crude oil production to refuelling the vehicle must also be considered. In such a *well-to-wheel consideration of* emissions, the advantage of electric vehicles then becomes abundantly clear.

The bottom line is that the emission behavior is the primary and decisive advantage of electric drives. Only with them will it be possible to sustainably comply with the ever stricter emission laws worldwide. Moreover, the absence of emissions can be experienced directly by the user and his environment: a car without a combustion engine and without an exhaust does not stink and does not make noise.

4.1.5.2   Compact Vehicle Concepts

Several factors lead to BEVs requiring far fewer components than internal combustion vehicles and therefore being designed much more compactly than vehicles with internal combustion engines, despite the relatively large battery:

- Electric motors have a significantly higher specific power than combustion engines, so they are much lighter and smaller with the same engine power. In terms of supply lines, electric motors only require a cable connection to the engine control unit and no fuel or coolant lines, which means that the desired drive power can also be very easily distributed to several small motors, which can then be accommodated in the wheels. With such *wheel hub motors,* a simple and flexible realization of front-wheel, rear-wheel, or all-wheel drives is possible even without differential, transfer case, and output shafts.

- Since the speed of the electric motor can be regulated continuously from standstill to maximum speed (which is significantly higher than that of combustion engines, e.g., 11,400 rpm in the BMW i3), there is no need for a manual gearbox. Reversing is made possible by simply reversing the polarity of the motor voltage – although the maximum speed that can then be achieved is of course limited.
- Since there is no combustion in the electric motor, there are neither exhaust gases nor heat-conducting components. There is therefore no need for an exhaust system, front radiator, or heat shields. In particular, no safety clearance to hot parts need to be maintained in the vehicle design.
- Last but not least, a battery placed flat in the underbody allows the realization of a flat vehicle floor without tunnels for cardan shafts or exhaust pipes – which in turn allows for a spacious and comfortable design of the interior.

The fact that this allows an electric vehicle to have a significantly larger interior than an internal combustion vehicle of the same size often leads to a surprised "It's much roomier than I thought" when getting into the vehicle for the first time. A BMW i3 BEV, for example, is significantly smaller than the BMW X1 compact SUV in terms of external dimensions but comparable to the larger BMW X3 mid-range model in terms of interior space.

The situation is completely different for PHEVs. Here, both an electric and an internal combustion engine powertrain must be accommodated, i.e., electric motor and combustion engine as well as battery and fuel tank, which inevitably leads to a reduction in the available space. For this reason, the space available in the rear row of seats or the boot volume is often reduced in PHEVs compared to a pure combustion engine variant.

4.1.5.3 Driving Dynamics and Agility

In the case of the internal combustion engine, the maximum torque is only available within a narrow speed range, which is why a multi-speed manual transmission and – depending on whether this transmission is shifted manually or automatically – a mechanical or hydraulic clutch between the engine and wheels are required for acceleration. In order to achieve the highest possible acceleration performance at the start, the engine must be loudly "revved up" in advance; the gearshifts required over the acceleration process then cause a brief reduction in engine power and the noticeable jerk.

In comparison, anyone driving an electric vehicle for the first time will be impressed first and foremost by the effortless acceleration. This is due on the one hand to the already mentioned higher specific power of electric motors but on the other hand also to the way they develop it: In contrast to the internal combustion engine, the electric motor has full torque available to it from standstill until maximum engine power is reached, resulting in a particularly strong and even increase in speed. While small electric vehicles in particular were ridiculed for a long time in terms of their driving performance, today they have left the competition of combustion vehicles far behind in terms of acceleration performance. A Tesla Model S P90D, for example, which is less of a sports car than a luxury sedan, has a maximum output of 396 kilowatts and thus accelerates from 0 to 100 kilometers per hour in about 3 seconds despite its unladen weight of 2200 kilograms.

The electric drive also scores in terms of agility. For one thing, the battery, as the heaviest component in the design, is usually positioned in the underfloor of the vehicle, giving electric vehicles a comparatively low center of gravity. This reduces the relative movements of the body in relation to the wheels, even in taller vehicles, and thus leads to noticeably good roadholding. On the other hand, electric vehicles have a very large steering angle of the front wheels and thus a very small turning circle (9.8 meters in the case of the BMW i3), which makes parking and turning in confined spaces noticeably easier and, in many cases, possible in the first place. In an internal combustion vehicle, on the other hand, the turning angle is generally limited by the engine mounts required to attach the engine-transmission unit, which run through the engine compartment as part of the body structure in the direction of travel and thus limit the depth of the wheel arches and the possible turning angle of the front wheels.

In fleet operation, where vehicles are often driven short distances and then parked again for some time, another advantage comes into play: electric vehicles do not require any operating temperature; they are immediately ready for operation even after a longer standstill, even up to full load. Also, the heating does not have to wait for the engine to warm up either.

4.1.5.4 Ride and Interior Comfort

Another advantage of electric vehicles is the quantum leap in driving comfort that can be directly experienced by drivers and passengers alike. Since there are no combustion processes in the electric drivetrain, no gearshifts and clutch operations are required, and there are no vibrating masses (such as the pistons in the combustion engine); hardly any vibrations are transmitted to the vehicle, especially at low speeds. On the one hand, this leads to a significantly higher level of acoustic comfort. If you ask owners of electric vehicles today what they would miss most if they had to switch back to a car with an internal combustion engine for range reasons, the experience of driving quietly is usually at the top of the list and is often compared with gliding along in a spaceship. On the other hand, the freedom from vibrations also contributes to this feeling, which also delays the vibration-induced loosening of clipped trim parts, for example, which otherwise leads to disturbing noises such as rattling and creaking in the interior with increasing mileage.

In combustion vehicles, an additional fuel-powered heater is required for the realization of a parking heater, which is integrated into the vehicle's heating/air conditioning system and usually has to be ordered as optional equipment or accessories. A parking cooler is not even available in the luxury class due to a lack of technical solutions. In the electric vehicle, on the other hand, both are available without any additional effort or cost: both the vehicle's heating and air conditioning are electrically operated. The power is supplied by the vehicle battery, so heating and cooling can also be provided when the vehicle is stationary. Ideally, the battery is connected to a charging station so that the required energy is not at the expense of the range. This ensures that the vehicle is fully charged at the time of departure, the interior is at a pleasant temperature, and the vehicle battery is also at an optimum temperature.

Anyone who travels a lot by car on business would often also like to work or rest in it for a short time. What makes this unattractive today, apart from the design of the interior, are the temperatures in the vehicle. It's too hot in summer and too cold in winter, and running the auxiliary heating or even the engine for heating or cooling is not an option. Here, too, the electric drive offers advantages: Because the vehicle interior can be heated and cooled silently and without emissions, it is also possible to stay in the stationary vehicle for longer periods at pleasant temperatures, regardless of the prevailing outside temperatures.

4.1.5.5 Operating Costs

In contrast to the abovementioned advantages in terms of emissions, vehicle design, dynamics, and interior comfort, the significantly lower operating costs of an electric vehicle compared to a combustion engine vehicle can be experienced less directly – but are no less attractive to the user for that reason.

If we take the usual market prices in Germany in 2019, the energy price of super petrol is 0.16 euros per kilowatt hour at a retail price of 1.35 euros per liter, and that of diesel is 0.13 euros at a retail price of 1.25 euros per liter – which is still roughly half as cheap as the energy price of electricity, which is 0.30 euros per kilowatt hour. However, what makes the electric drive cheaper than the combustion engine in terms of energy costs is the significantly higher efficiency, which is around 30 percent for the gasoline engine, around 40 percent for the diesel engine, and over 90 percent for the electric motor. In addition, it is possible to recover up to 20 percent of the energy during braking, especially in urban traffic, by means of recuperation, which in the case of the internal combustion engine is converted into heat by the brake and thus lost.

In addition, electricity costs can be further reduced by integrating the vehicle into a V2G network, as described above, or avoided altogether by using electricity generated by photovoltaic systems, for example. Also, even if – as is foreseeable – charging electricity is one day taxed at a higher rate, similar to petrol and diesel fuel, the energy costs of electric vehicles will remain more favorable in the event of a further price increase, which is also foreseeable.

The significantly lower service requirements of the electric powertrain also have a positive effect on operating costs: Typical wear parts such as clutch linings, piston rings, timing belts, spark plugs, water pumps, or all types of filters do not exist, and the often costly service measures associated with their replacement are just as unnecessary as oil changes. The ability to brake the vehicle during normal driving solely via recuperation also significantly delays brake pad wear. In addition, the smooth running already mentioned above has a positive effect on the service life of mechanical and electronic vehicle components. The vehicle batteries, initially often seen as a risk factor, have a significantly longer service life than originally assumed and do not play a role when considering operating costs in the first few years.

Electric vehicles are of course superior to vehicles with combustion engines primarily because they are locally emission-free – but also because of their compact design and their significant plus in driving dynamics and comfort. Anyone who wants to be convinced of the advantages of an electric vehicle must experience them.

4.2 Autonomous Driving

From the princes of the Orient who fly elegantly to their destination on flying carpets in the stories of the Thousand and One Nights to the families happily playing cards in their cars, whose vehicles move all by themselves on elevated roads through futuristic cities in the future images of the 1960s, the vision of the autonomous, or driverless, vehicle has always fascinated people. However, what was still a fairytale and a dream back then is now on the verge of becoming a reality. Today, the necessary technologies have reached a level of maturity that makes the implementation of autonomous driving within reach. At the same time, however, it is precisely the fascination that attaches to the topic that leads to the hype; not a day goes by without reports and speculation in the media about further progress, the imminent breakthrough, and, above all, the enormous business potential.

In order to be able to realistically assess the role that autonomous vehicles will actually play in the future, one cannot avoid first dealing with the technical fundamentals: What must such a vehicle be able to do, where do the technologies required for implementation stand, and which driving situations, however rare, must be taken into account and safeguarded during development? In addition, the legal framework must be considered: Which technology is approvable in which market? What legal and thus financial risks are taken by companies or individuals who manufacture, operate, or use autonomous vehicles? Last but not least – as fascinating as the idea and the technology may be – who would really want to be driven by an autonomous vehicle in the end?

4.2.1 Technology

4.2.1.1 Levels of Automation

In everyday language, we speak of assisted, autonomous, automated, or piloted driving, self-driving vehicles, or robo-cars. In order to create clarity here, the German Federal Highway Research Institute (BASt) proposed five different levels of automation as a nomenclature back in January 2012. Based on this, the American standards institute SAE International adopted the SAE J 3016 standard in 2014, which is now used by most automotive manufacturers. Compared to the five levels of the BASt proposal, SAE J 3016 defines six levels of vehicle automation; *conditionally automated driving* is differentiated here in *partially automated driving* and *highly automated driving:*

- Stage/Level 0: No automation (*"driver only"*):
  The driver moves the vehicle completely independently. Support systems such as ABS, high beam assistant, and lane departure warning assist the driver but do not actively intervene in vehicle control at any time.

- Stage/Level 1: Assisted driving (*"feet or hands off"*):
  Assistance systems such as adaptive cruise control (ACC) or lane centering assistant (LCA) support the driver in steering the vehicle longitudinally or laterally by actively intervening in the cruise control or steering. However, the driver retains full control at all times.
- Stage/Level 2: Partially automated/partially autonomous driving (*"feet and hands off"*):
  Both longitudinal and lateral control of the vehicle can be taken over by assistance systems in certain driving situations – for example, by a traffic jam assistant in a traffic jam or by a parking assistant when parking. Here too, however, the driver must have full control of the vehicle at all times.
- Stage/Level 3: Conditionally automated/conditionally autonomous driving (*"eyes off"*):
  Longitudinal and lateral guidance of the vehicle can be completely taken over by assistance systems over longer periods of time – for example, in traffic jams by a traffic jam chauffeur. The vehicle continuously monitors its surroundings with the aid of its sensors. When prompted, the driver must be able to take over control of the vehicle again within 10 seconds.
- Stage/Level 4: Highly automated/autonomous driving (*"brain off"*):
  The vehicle is guided in a fully automated manner and can, for example, park in a multistory car park on its own. Although a driver is required to take over the driving of the vehicle again on request or at his or her own request, he or she is no longer in charge. Level 4 driving can also be enabled only for partial areas, such as the motorway or city traffic.
- Stage/Level 5: Fully automated/fully autonomous driving (*"driver off"*):
  The vehicle is fully automated under all driving and environmental conditions; driver intervention is neither intended nor possible. A Level 5 vehicle differs from a Level 4 vehicle primarily in that a driver's workplace with the associated possibilities for control and operation no longer exists.

Levels 0 to 2 have in common that, despite support, the driver must be in control of the vehicle at all times and must observe the vehicle's surroundings for this purpose. Only from level 3 onward does the vehicle monitor its surroundings independently and can therefore take over the driving of the vehicle itself for a certain period of time, which varies depending on the level.

Level 1 systems are already in widespread use today; Level 2 systems are available in the latest models from premium manufacturers. The introduction of Level 3 vehicles, which will then be able to change lanes independently, for example, was announced for the end of 2018 but has been postponed several times for legal reasons. Highly and fully automated vehicles should then be available between 2021 and 2025, respectively.

4.2.1.2   System Requirements

Function

The primary task of a system for autonomous vehicle control is the implementation of driving orders: To do this, the user first enters the destination of their journey, for example, via voice recognition or app. The first thing the vehicle does is to check whether it can carry out the desired journey at all, whether there possibly are too many people on board, or whether the battery charge level is insufficient for the journey. Based on its current position, the vehicle then uses a highly accurate digital map and additional information such as the current and forecast traffic situation to determine the optimal route to the destination. This target route is then implemented by the system via the corresponding control of the actuators, which are primarily the engine, brakes, and steering but also lights or turn signals. During the journey, the vehicle's position is recorded via GPS, camera-based recognition of road markings, special positioning signs at the roadside, or measurement of the vehicle's movement and corrected if necessary.

Far more complex than the mere implementation of the driving task is the second task of autonomous vehicle control, namely, to react adequately to the vehicle environment at all times and thus to ensure the safety of the vehicle passengers and the environment. To do this, the system must continuously monitor the vehicle environment and reliably detect situations that require a response from the vehicle. This may involve the behavior of other road users; signals such as traffic lights, traffic signs, or road markings; or special traffic situations such as road works or black ice. In any case, the system must comprehend the situation holistically, make the right decision, and then implement it immediately – at any time and extremely quickly: A system for autonomous vehicle control must be real-time capable and enable the entire process from the occurrence of an event, through its detection by a sensor, the evaluation of the sensor data, and decision-making, to the physical reaction of the vehicle within a few milliseconds.

Safety in Use

The "control and reduction of risks and hazards in the case of intended use and foreseeable misuse", which is referred to as *safety in use*, is the technical challenge par excellence in the development of autonomous vehicles. In principle, the mechatronic systems on which the vehicle functions are based must function under all expected conditions, as in conventional vehicles, i.e., also on all conceivable surfaces, in a wide variety of weather conditions or under the influence of different external electromagnetic fields. In particular, the protection of the system against unauthorized access to or manipulation of vehicle data is also part of the basic safety of use.

The particular challenge of autonomous vehicle control is that the situation analysis and subsequent decision-making must not give rise to any unacceptable risks for vehicle users and other road users. The underlying algorithms and knowledge bases must therefore offer maximum safety. At the same time, however, this is also where the greatest potential of autonomous vehicles lies: Once a system for controlling an autonomous vehicle has learned to interpret the behavior of its environment correctly and to draw the right conclusions from it, it can operate much more safely than a human driver, whose attention and perception are subject to fluctuations and are always limited in some way, due to its constant unrestricted attention, its 360° all-round view, and the possibility of detecting circumstances that are not yet perceptible to humans (such as an oil puddle located in a curve that is difficult to see but reported by another vehicle).

In addition, the safety of use of an autonomous vehicle includes knowledge of its own functional limits, such as when driving through a complex construction site or when driving in heavy snowfall. Such special situations, which the system alone may not be able to handle, must be detected by the vehicle in good time so that it can either bring itself to a safe state or – in the case of Level 3 and 4 systems – hand over the driving task in good time to the driver who is standing by.

Functional Safety

In addition to the safety of use described above, the *functional safety of* a system includes the prevention of risks and hazards resulting from a malfunction of this system. Since in the case of autonomous control of motor vehicles a system error quite obviously leads directly to a serious hazard for the vehicle user or other road users, the requirements with regard to functional safety for autonomous vehicles are accordingly extremely high. In the automotive industry, the risks posed by technical systems are assessed and classified on the basis of their *Automotive Safety Integrity Level (ASIL)*. Three criteria are included in the calculation of the ASIL:

- The severity of the consequences of a possible error:
  This is extremely high in an autonomous vehicle control system; an unintended steering movement, for example, can steer the vehicle into incoming traffic and thus quickly lead to serious injuries or the death of one or more people.
- The frequency of use of the system:
  A system for autonomous vehicle control is not only in use from time to time, but permanently, so a fault can occur at any time of vehicle use and can lead to a hazard.
- The ability to mitigate the effect of the fault by external intervention:
  These are very limited in autonomous driving; especially in Level 4 and Level 5 vehicles, the passenger(s) have little opportunity to take corrective action if the system makes a mistake.

Systems for autonomous vehicle control therefore achieve a very high ASIL and are accordingly classified in the highest risk group *ASIL D*. ASIL D systems require the highest level of reliability, with a maximum of one fault permitted in 100 million hours of operation. So if 10,000 autonomous vehicles drive for 10,000 hours each, a maximum of one fault may occur in total.

4.2.1.3 System Architecture and Components

A system for autonomous vehicle control that is intended to meet the abovementioned functional and safety requirements first needs appropriately safe subsystems and components such as sensors for environment detection, drive, brake, and steering systems or even digital road maps. Beyond this, however, what matters is its higher-level structure and its integration into the *system architecture of* the overall vehicle, i.e., into the overall structure of all the electrical and electronic subsystems belonging to the vehicle – from the engine and transmission control system and the entire infotainment system to the windscreen wipers, horn, and lighting.

System Architecture

Vehicle functions are controlled via *control units*, i.e., "mini-computers" of different sizes and different functional and performance ranges installed in the vehicle. Each control unit controls the functions of one or more vehicle components – from the windscreen wiper to the engine. Today's passenger cars usually have a highly decentralized system architecture with many small control units distributed throughout the vehicle.

In contrast, the system architecture of autonomous vehicles has a large, central control unit that takes over the sensitive driving and monitoring tasks. Extreme demands are therefore placed on such a control unit in terms of performance and reliability, and the development and manufacturing effort involved is considerable. At the same time, for reasons of weight and cost, the many small control units in the other, classic functional domains such as infotainment (navigation, telephony, entertainment, etc.), body electronics (lighting, window cleaning, window lifters, etc.), or safety (airbags, seatbelt tensioners, etc.) are being replaced by more powerful central domain control units. There is a clear trend toward centralization in system architecture development.

The components required for autonomous driving, such as sensors (cameras, motion sensors, etc.) and actuators (engine, steering gear, brake, etc.), are connected to this central control unit via an *in-vehicle network*. In addition, the other domain control units and one or more antennas are connected to this in-vehicle network for permanent data exchange with the *backend*, a central server that performs computing operations with high power requirements and realizes the data exchange between vehicles and their digital environment. The high functional and safety requirements for autonomous driving necessitate the use of extremely fast and reliable network technology. The solutions currently used in vehicle technology for data transmission, such as CAN, LIN, or FlexRay, are not suitable for this purpose, so the Ethernet standard, which has been tested and established in home and office computing, is used. Today, Ethernet already offers a data rate of 10 megabits per second; in the future, up to 100 gigabits per second will be possible.

Connectivity

In addition to the connection of the vehicle's internal components, autonomous driving and the realization of other vehicle functions and services also require the corresponding *connectivity*, i.e., the fast and secure exchange of data between the vehicle and its digital environment. First and foremost, there is the connection to the aforementioned backend, referred to as *vehicle to backend (V2B)*. The backend is a central server operated by the service provider responsible for the function and safety of the vehicle, usually the vehicle manufacturer. A reliable and secure connection of the vehicle to the Internet via the backend is the basis of essential vehicle functions:

- The transmission of relevant environmental information such as climate data, road conditions, or the position of free parking spaces for use in *location-based services (LBS)*. Any communication with servers of external service providers in this context also runs via the backend for security reasons.
- The installation of new software releases *Remote Software Updates (RSU)* to improve or extend the functionality of vehicles.
- The possibility of remotely activating or deactivating functions provided in the vehicle's hardware creates new pay-on-demand business models for operators. For example, sunroofs can be installed in all vehicles of a car sharing fleet, but the possibility to use them can be linked to the payment of a fee.
- By collecting usage data, information on customer behavior and vehicle condition is generated, from which customer-oriented products and service offers can be derived.

The technical prerequisites for V2B data exchange are one or more antennas connected to the central control unit on the vehicle side and a fast and reliable mobile Internet connection. This is especially true for autonomous vehicles, which must be able to exchange video data of the vehicle environment securely and in real time with the backend, for example. The data transmission rate, reliability, and coverage of the 4G/LTE mobile communications standard available today do not meet the necessary requirements here; autonomous driving requires the future 5G standard, which has a bandwidth of up to 10 gigabits per second and a latency of <1 millisecond. However, nationwide availability of 5G will still take years; in Germany, it is planned by 2025. So if autonomous vehicles can only drive where 5G is safely available, the introduction of 5G is one of the limiting factors for the introduction of autonomous driving.

Beyond the connection to the vehicle backend, an autonomous vehicle maintains other wireless communication links with its digital environment:

- Vehicle to Infrastructure (V2I):
  By exchanging data with intelligent traffic infrastructure, the vehicle can, for example, be informed by a traffic light of the time remaining until the next signal change, which enables anticipatory, safe, and energy-saving driving. In the same way, the payment process for road tolls, parking, refueling, or charging can be handled directly with the provider's infrastructure.
- Vehicle to Vehicle (V2V):
  The direct exchange of data with vehicles in the immediate vicinity enables potential hazards to be identified or useful information to be passed on, such as the condition of the road or obstacles on the road.
- Vehicle to User (V2U):
  Data exchange with users inside and outside the vehicle enables, for example, the use of addresses available on a smartphone as a destination, access to personal music libraries, or the quick opening of the vehicle via smartphone without the detour via the backend.

Sensors

Even conventional vehicles today already have a whole range of sensors that provide the input information for the assistance systems:

- Wheel speed sensors, e.g., for the anti-lock braking system (ABS)
- Position sensors, e.g., for vehicle dynamics control (ESP)
- Distance sensors for close range, especially for Park Distance Control (PDC)
- Camera systems, e.g., as rear view camera, for traffic sign recognition or the recognition of lane markings
- Radar systems for distance measurement in the long range, as required, e.g., for active cruise control (ACC)

For automated driving from Level 3, however, the performance spectrum of these sensors is no longer sufficient; here, comprehensive detection of the vehicle environment up to a distance of 300 meters is required. Both stationary and moving objects such as pedestrians, cyclists, vehicles, road markings, curbs, or bridge piers must be detected here – and under all possible ambient conditions such as rain, fog, or snowfall. For this purpose, the so-called *light detection and ranging (Lidar)* systems are used today, which map the vehicle environment at high speed as a 3D point cloud, from which 3D objects can then be created in real time and identified using stored, learned knowledge.

Autonomous test vehicles in use today – from BMW, Tesla, or Waymo, for example – are clearly recognizable as such from afar by the large lidar systems mounted on their roofs. These work by means of lasers and rapidly rotating mirrors whose light is reflected by objects in the vicinity and then picked up by a sensor in the lidar. Today, such rotating laser systems can record up to 1000 images per second, but they cost up to 70,000 euros or more – which in many cases is twice the price of the vehicle below and would thus quite obviously be a killer criterion for the spread of autonomous driving.

Solid-state lidar is a promising approach for systems that are significantly cheaper to manufacture and at the same time more robust in use. Although their observation angle is limited to about 120°, they do not require the expensive and wear-prone rotational mechanics.

At the same time, the sensor systems must be designed with multiple redundancies; i.e., several identical systems must always be available at the same time – in order to meet the aforementioned ASIL-D requirements for functional safety. In order to make rapid progress in the development of this key technology, which is essential for autonomous driving, despite such high demands, the classic Tier 1 suppliers to the automotive industry are cooperating here with the lidar system specialists, such as Continental with ASC, ZF with Ibeo, Infineon with Innoluce, Magna with Innoviz, or DENSO with Trillumina. As a result of these intensive and broad-based development activities, it is expected that lidar systems will be available on the market at a price of <100 euros by 2025.

Actuators

The primary control tasks of a vehicle include steering, acceleration, and deceleration, and the secondary tasks include setting signals or opening and locking windows, flaps, or doors. The actuators required to perform these control tasks are basically available today and also largely meet the requirements for real-time capability and reliability: electronically controlled steering actuators from steer-by-wire systems, electronically controlled brake actuation from ABS, and electronic engine control from active cruise control systems.

For the traction motors of autonomous vehicles, electric drives have the great advantage over internal combustion engines in that the sensitive electronic control of speed and torque required for safe and comfortable operation is technically much easier to implement with electric motors than with internal combustion engines.

Digital Maps

Digital maps available today were developed for use in navigation systems and are sufficient for this purpose; however, they do not meet the high requirements of autonomous driving in terms of accuracy, up-to-dateness, and coverage. What is needed here are the so-called *high-definition (HD) maps* that depict the road network and traffic infrastructure in a highly accurate 3D model. In order for an autonomous vehicle to be able to move safely on the basis of the HD map, it must be possible to distinguish individual lanes on HD maps. At the same time, the third dimension must also be taken into account so that the vehicle knows, for example, exactly on which floor of a parking garage it is currently located.

In addition to accuracy, the main challenge is the required real-time up-to-dateness: For example, a road must be marked as closed on the map more or less the moment the worker stretches the barrier tape across the road. The required accuracy and topicality of HD maps is achieved by *mobile mapping*, i.e., the continuous recording of roads by vehicles, aircraft, or drones equipped with special cameras, as well as by the permanent updating of maps on the basis of images of the vehicle's surroundings, which are recorded by the vehicle's own cameras and fed into the HD map via the backend. Among other things, relevant dimensions such as the width of the road or the clearance height of an underpass are also recorded.

Localization, i.e., determining the exact location of the vehicle, is also based on the HD map. For this purpose, the image of the surroundings captured by the vehicle sensors is continuously compared with the HD map, which allows the position to be determined with an accuracy of <50 centimeters in the direction of travel and <15 centimeters to the right and left.

In addition to this geometric information, digital maps contain a number of other details required for autonomous vehicle guidance. These include, first and foremost, current speed limits, turn restrictions, or parking rules.

The creation and maintenance of accurate and up-to-date digital maps involve extremely high and permanent costs, while at the same time their widespread availability is a necessary prerequisite for the feasibility of autonomous driving. In order to spread the financing of these costs over several shoulders and at the same time to implement common standards for HD maps, consortia consisting of representatives of the automotive industry and digital players were formed early on. Leading providers of digital HD maps in Europe and North America today are Google, HERE (owned by Audi, BMW, Bosch, GIC, Intel, Continental, Mercedes, NavInfo, and Tencent), and TomTom, as well as Baidu/ZF in China.

Learned Behavior and Artificial Intelligence (AI)

In order to drive a vehicle safely for all road users and as comfortably and quickly as possible for the vehicle occupants, a driver unconsciously makes predictions about the behavior of other road users based on the environmental data he or she has recorded and compared it with the current state of movement of his or her vehicle, derives his or her own options for action, makes appropriate decisions, and translates these into corresponding actions. For experienced drivers, this usually takes place in an anticipatory, continuous, and unconscious manner.

This is exactly how an autonomous vehicle should behave. However, such a system behavior cannot be written as a program in the traditional way and stored in the control unit; the vehicle must first learn it through *deep learning*, a method of artificial intelligence (AI). To do this, the autonomous vehicle is first "shown" a large number of objects that it detects via its sensors. For each object, the vehicle is then "told" what it is – a cyclist, a pedestrian, a bridge pillar, or whatever. In order to teach the vehicle as much knowledge as possible as quickly as possible, it is shown not only real objects but also images and simulated driving situations. In this way, the vehicle learns to reliably recognize identical objects and to distinguish them from others.

Once the vehicle is familiar with the objects in its world, it is then trained in a second step to behave appropriately in each case, such as stopping, swerving, accelerating, or simply driving on. Again, this is not primarily done in real environments but through images, films, and simulations. Through this training – analogous to the human brain – artificial neuronal networks are mapped by means of a suitable program, which establish associations between recognized objects or situations and derive decision-based actions from them.

Before an autonomous vehicle is then allowed to participate in road traffic on its own responsibility – i.e., in Level 4 or Level 5 operation – its neural network must be sufficiently trained in this way. Once this has been achieved, it continues to learn by comparing the images recorded in its environment with a knowledge store accessible via the backend.

One possibly disadvantageous property of such neural networks is that – in contrast to classical program code – the behavior based on them cannot be reconstructed ex post. Why a vehicle reacted in a certain way in a certain situation can therefore no longer be determined in retrospect – a highly unsatisfactory situation from a legal perspective, for example, when determining the cause of an accident.

Autonomous Driving Platform

Processing the data recorded by the sensors, including self-localization, updating the digital environment model, predicting the behavior of other road users, and making decisions for the safe control of the vehicle require the on-board availability of a computing power that until now has only been available from stationary supercomputers. In order to be able to meet the aforementioned functional safety requirements, the central control unit used, the so-called *autonomous driving platform*, must have multiple redundant processors and structures.

One example of such a platform is the Nvidia Drive AV jointly developed by Nvidia and Continental. The Xavier processor used in it has more than 9 billion transistors, with which it can perform 30 trillion computing operations per second. Other platforms include ZF ProAI, also developed by ZF with Invidia, the Central Sensing Localization and Planning Platform from Delphi and Mobileye, and Apollo from Infineon and Baidu. It is worth noting that an established automotive supplier and an IT specialist always work together here. One or more car manufacturers are then also involved in the joint system development.

Just like the availability of the 5G network, the development and industrialization of low-cost and robust lidar systems for environment detection is clearly on the critical path for the introduction of autonomous driving.

4.2.2   Autonomous Vehicles

4.2.2.1   Vehicle Concepts

Up to and including Level 4, automated driving systems are, strictly speaking, high-performance driver assistance systems that support and relieve the driver in controlling the vehicle when needed – but only when needed. As long as he wants, however, he can drive the vehicle himself as usual, which is why proven vehicle concepts are also retained: The vehicles continue to have a "driver's workstation" from which he steers and operates the vehicle. The additional components required for automated vehicle guidance, such as sensors and control units, are integrated geometrically and functionally into the existing vehicle concepts.

Level 4 vehicles play a special role here, as they must be able to be steered through road traffic both as conventional vehicles by drivers and in autonomous mode by an autonomous vehicle control system. Similar to the comparatively expensive and heavy plug-in hybrid, in which two different propulsion systems must be kept on hand, a Level 4 vehicle must keep the systems for two alternative modes of control. They are burdened with the cost, space, and weight of the components required for both modes and thus ultimately represent an expensive compromise. The ability shown in many concept vehicles to retract the steering wheel and pedals in autonomous mode, allowing better use of the interior space in this situation, is an example of this duplication of requirements.

The situation is completely different for Level 5 vehicles, which are operated exclusively autonomously. Here, there is no longer any need to provide a driver's workplace, which has two far-reaching consequences for vehicle design: On the one hand, the elimination of the driver's workstation opens up completely new design possibilities for the interior. On the other hand, the fact that the vehicle can now be designed exclusively to meet the requirements of passengers opens up much greater scope and changed priorities in the design of overall vehicle concepts: there is no longer a driver who has personal requirements in terms of dynamics, agility, or driving pleasure, who takes risks in order to overtake someone quickly or forgets to set the indicator when turning. The focus of a Level 5 vehicle concept can thus be on comfort, connectivity, and entertainment, often summarized as *drive time utilization*. More or less radical visions and studies of such fully autonomous vehicles with seats facing each other and a lounge atmosphere are regularly shown at motor shows and in trade journals. However, these studies still pay little attention to the unresolved issues of passive safety and kinetosis – the discomfort caused to passengers by vehicle movement. Section 5.3.3 goes into detail on the requirements for vehicles to become dangerous.

A next stage in this sense would then be traffic systems in which only autonomous vehicles are on the road. Based on the premise that autonomous vehicles do not make any mistakes and therefore do not cause any accidents, there would no longer be any accidents between vehicles in such a system. In this case, all passive safety functions could be dispensed with on the vehicle side – which is not yet possible in mixed operation with conventional vehicles.

4.2.2.2 Demand, Requirements, and Acceptance

Autonomous driving fascinates and is more ubiquitous in the media than almost any other technology topic. On the one hand, of course, there are the spectacular reports about accidents during testing on public roads, unfortunately often with fatal outcomes. On the other hand, further back in the business news, the success stories about new partnerships, successful tests, newly developed high-performance sensors, or newly opened research centers predominate. At the same time, car manufacturers, suppliers, and system partners are investing so much money in the development of autonomous driving that one hardly dares to ask the real question behind all this: who really wants to use or operate such autonomous vehicles in the end?

However, especially when so much money is being invested in this technology, a well-founded answer to this question is essential. In order to be able to do this, it is first necessary to determine the actual advantages of autonomous driving compared to conventional driving with a driver. Also, when making this comparison, it quickly becomes clear that many of the options mentioned in the media that autonomous driving is supposed to open up, such as being able to sleep, work, or be entertained while driving, having one's vehicle driven into a parking garage, or even being mobile without being able to drive a vehicle (such as children, seniors, or the disabled), are not advantages of autonomous driving over driving with a driver but advantages of getting driven over driving oneself. As such, on closer inspection, they are completely independent of whether the passenger is driven by a human driver or by an automated machine.

At the end of a differentiated consideration in this sense, there remain exactly two relevant aspects from the customer's point of view, with regard to which an autonomously controlled vehicle that will be available at some point in the future could potentially stand out positively from conventionally driven vehicles: safety and costs.

- Improved safety:
  According to a study by the US National Highway Traffic Safety Administration (NHTSA), almost 95 percent of all traffic accidents in the USA today are caused by human error. Autonomous vehicles are – at least in theory – always attentive, adhere 100 percent to traffic rules, do not take any risks, do not overestimate themselves, and, on top of that, predict the behavior of other road users based on additional information and learned knowledge much better than even experienced drivers can. The potential for a quantum leap in road safety is therefore absolutely there. However, there is still a long way to go before this safety can be technically implemented. The mixed operation of autonomous and conventional vehicles described in the previous chapter is particularly critical and, realistically speaking, will be unavoidable, at least for a transitional period of several years.
- Lower operating costs:
  As in any business sector, the automation of recurring manual operations also offers business opportunities in mobility services. Today, providers have to find, employ, and pay drivers. They bear the financial risk if drivers fail to show up for duty for whatever reason or are involved in an accident with the vehicle. The use of autonomous vehicles therefore not only represents a significant cost-cutting potential for operators but also mitigates the economic risks. The decisive factor here is the point in time at which autonomous vehicles are not only technically reliable but are also available at a price that makes their operation in fleet use economical.

In principle, the uncertainty as to when autonomous vehicles will have reached the technical maturity required for these advantages to become a reality means that both potential customers and providers, as well as the municipalities concerned, view autonomous driving not only with acceptance or enthusiasm but also with scepticism or even rejection. When answering the abovementioned question of who actually wants autonomous cars, the very different goals of these three groups of stakeholders must be considered in a differentiated manner.

The View of Mobility Customers

Mobility customers who want to get from A to B by car as quickly, safely, comfortably, and inexpensively as possible usually have up to four alternative options to choose from. The decision is then made on the basis of individual requirements and current conditions:

- Driving yourself in your own car. The use of a privately owned car is still the standard option in the vast majority of cases.
- Driving yourself in someone else's car. This includes the use of a rental car or a car sharing service.
- Getting driven in your own car. This is possible either with a paid chauffeur or through autonomous vehicle control.
- Getting driven in someone else's car. This includes all ride-sharing services such as taxis or Uber, and the vehicle can be controlled by a driver or an autonomous vehicle control system.

Being able to have your own car drive autonomously when needed (i.e., in Level 3 or 4) is a costly but also highly attractive comfort function. Particularly in city traffic or on the motorway, there are driving situations in which the driver is happy to delegate responsibility, hand over control of the vehicle to the autonomous system, and use the time otherwise. The possibility of being able not only to make phone calls but also to read or write while driving, and thus to really use driving time as working time, makes an optional autonomous driving system particularly attractive for company cars. Due to the high costs associated with the system, demand for such optional equipment will primarily be for vehicles in the upper price segment.

The situation is completely different with Level 5 vehicles, which no longer have a driver. It is highly unlikely that private individuals will purchase such fully automated vehicles on a broad basis; at most, market niches will open up here. In the luxury segment, for example, a kind of autonomous "private jet on wheels" is conceivable, but it is precisely here that a driver is perceived as a significantly higher-value alternative to an autonomous driving system – the associated costs play a much smaller role in this segment. Another possibility could be small autonomous city vehicles for active seniors who no longer drive themselves but still want to retain their mobility in their own vehicle. The potential of fully automated vehicles, on the other hand, is significantly lower when it comes to transporting people in need of assistance, such as children, frail seniors, or the disabled, which is frequently mentioned as an advantage. The care and safety function provided by the presence of a driver, which an autonomous driving system by its very nature cannot provide, is clearly of particular importance in these cases.·

The fundamental question of the acceptance of mobility services – whether in an autonomous or driver-guided vehicle – versus the use of a personal vehicle is discussed in detail in Chap. 5.

View of the Municipalities

While national and state authorities are primarily interested in the economic potential of autonomous driving, cities and municipalities are directly affected by the possible use of autonomous vehicles. They have to solve the concrete traffic-related problems of their area of responsibility by means of regulations and laws on behalf of the citizens and thus their voters. When it comes to the pros and cons of autonomous vehicles, three aspects are at the forefront:

- Safety: Can the use of autonomous vehicles really reduce the number of traffic fatalities and injuries, or does it not also entail new safety risks for passengers or passersby?
- Traffic flow: Will the anticipatory driving of autonomous vehicles lead to smoother inner-city traffic, or will morning and evening traffic jams get worse because autonomous vehicles drive over-cautiously and slowly?
- Quality of life: Do autonomous vehicles, even if powered by an internal combustion engine, lead to a reduction in harmful emissions – and does the need for public space for driving and parking private cars decrease?

Today, many cities are faced with the concrete decision of whether to ban the operation of autonomous vehicles on their roads, allow it under conditions that have yet to be agreed, or even promote it. Initially, this involves pilot operation with a few vehicles that are not yet 100 percent technically mature, with which their effects on the traffic of the respective city are to be investigated. Only at a later stage will it be possible to introduce autonomous taxis across the board, for example. Based on previous experience, the argumentation can be summarized as follows:

- The main argument in favor of promoting autonomous vehicles is the aforementioned potential for improving road safety; the driver's inattention, which is the cause of the majority of traffic accidents, is eliminated here. Another advantage from a municipal perspective is the possibility of integrating autonomous vehicles directly into an urban traffic management system and thus optimizing the overall traffic flow. The promotion of autonomous driving benefits local companies and sends a clear signal about the innovativeness of a city, which in turn attracts other companies and investors.
- Reasons against autonomous vehicles in the city include the safety risks in the introduction phase. Even if the autonomous vehicles are fully developed and drive safely, the unavoidable mixed operation in the introductory phase in particular harbors risks. In addition, in many places, not only in Chinese cities, the authorities want to have a responsible driver on site in the event of an accident and to be able to hold him or her accountable. The critical question of whether the offer of low-cost mobility services based on autonomous taxis will not draw more users away from public transport and onto the roads, thus further worsening the traffic flow there, must be answered for each city within the framework of pilot projects. Another aspect to be clarified from the perspective of the municipalities is the threat of job losses for taxi drivers due to the then declining competitiveness of classic taxi services.

Against the backdrop of this diversity of advantages and disadvantages, there is no clear-cut opinion in very few metropolises today. Autonomous driving is only really promoted in municipalities where companies involved in its development are at home, for example, in Silicon Valley, Arizona, or Chongqing. In view of the increasingly critical traffic conditions, however, the primary focus of the cities is on reducing the number of vehicles in the city (cf. Sect. 6.4.2). Whether autonomous vehicles can really make a contribution here clearly still needs to be confirmed.

View of the Mobility Providers

In the narrow solution space remaining between customer wishes and regulatory requirements, the providers of mobility services want to place a business model that is as successful as possible. This can be self-driving services, such as car rental or car sharing, as well as services for getting driven, such as ride hailing, taxi, or ride sharing.

Autonomous vehicles make little sense as rental cars or in car sharing. On the one hand, they are very expensive to purchase due to the complex Level 3 or Level 4 technology, and on the other hand, mobility customers use car-sharing services precisely when they want to drive themselves and would otherwise be able to switch to ride-sharing services at any time. Also, car sharing with fully autonomous Level 5 vehicles is by definition ride-hailing, because the customer does not drive himself but gets driven.

From a mobility provider's point of view, the greatest potential of autonomous vehicles lies in the potential savings in driver costs. This is one of the reasons why major ride-hailing providers such as Lyft, Uber, or Didi have become massively involved in the development of autonomous vehicle controls. Driverless *robo-cabs* not only provide the actual transport service autonomously by picking up customers directly at their location – requested via app – and then taking them to their desired destination alone or together with other passengers but also park, refuel, or charge autonomously and distribute themselves in the city in a way that is optimal for users. This means that they are not only an alternative to the user's own vehicle and a supplement to local public transport but also – depending on the price of the vehicle – a highly attractive business model for the provider.

Whether mobility customers will also accept robo-cabs in the premium segment is, however, more than questionable. Here, in addition to a comfortable interior and adequate connectivity, help with loading luggage, a restaurant recommendation if needed, or perhaps just a nice chat during the ride are expected. For such additional services, however, a driver is required for whom the customer must be willing to pay.

What is being discussed today among experts and the public as the advantages of autonomous driving over conventional driving are primarily the advantages of getting driven over driving yourself. I think it is highly unlikely that private individuals will acquire driverless vehicles.

4.2.3 Legal Aspects

Safe technical implementation in the vehicle is, of course, the main challenge on the road to autonomous driving. However, the establishment of a clear legal framework is just as important for the development – here specifically the testing and ultimately the operation of the vehicles. Without the necessary legal certainty for manufacturers, operators, and municipalities, autonomous vehicles will not move on public roads anywhere in the world.

For the rapid and successful introduction of autonomous driving desired from the manufacturers' point of view, an equally rapid implementation of internationally uniform legal regulations is extremely important. In addition to the administrative law question of the requirements for licensability, the necessary legal framework must above all also clarify the civil or criminal law question of liability in the event that the operation of an autonomous vehicle causes damage to an object or person.

4.2.3.1 Approval Requirements

An essential legal prerequisite for the registration of motor vehicles in Europe is compliance with the *Vienna Convention on Road Traffic, which* was adopted at a UN conference in 1968 and has been ratified by over 70 countries to date. Article 8 of this convention explicitly requires that every vehicle in motion must have a driver and that the driver must be in control of the vehicle at all times – which obviously fundamentally rules out the eligibility of autonomous vehicles for registration. In September 2016, Article 8 of the Agreement was amended to allow the use of highly and fully automated vehicles (i.e., Levels 4 and 5) in the future, provided there is a driver on board who can override the systems if necessary. However, truly autonomous driving without a driver is still not permitted.

Since the USA and China have not ratified the Vienna Convention, the legal framework in favor of autonomous driving could be implemented there much more quickly. Companies such as Alphabet Waymo or Uber have already been testing robo-cabs in Arizona for some time. Despite several accidents with autonomous test vehicles, some of which were fatal, California has even allowed Level 5 vehicles to be tested without drivers since 2016, as long as there is a constant connection to the operator and a maximum speed of 35 miles per hour is not exceeded. In China, expected to be the largest market for autonomous vehicles, approvals are much more restrictive by comparison. Here, test drives have only been possible since 2016 and only on selected driving routes and only on the condition that there is a driver in the vehicle who can take control at any time. In addition to Baidu and Tencent, Daimler and BMW are also conducting autonomous test drives in China.

The primary prerequisite for an autonomous vehicle's eligibility for approval is in any case the prior demonstration of its functional safety. To this end, the requirements that an autonomous vehicle must meet must first be agreed on the basis of a set of scenarios (e.g., "child runs in front of the car" or "construction site drive-through"). In a next step, it is then necessary to define and agree on the scope of testing that must be carried out by the manufacturer to ensure that these requirements are met. The scope of testing required to validate all scenarios for the approval of an autonomous vehicle can amount to several hundred million kilometers, which is why a large proportion of these tests are not carried out on roads but on test benches or by simulation (virtual vehicle in a virtual environment). Alphabet Waymo, for example, has conducted over 13 million kilometers of testing on public roads in Arizona since 2009 and is still only allowed to operate its autonomous vehicles there under restrictive conditions.

Other aspects relevant to registration that are under discussion include a possible restriction of the roads on which the vehicle may drive, the specification of a maximum speed and maximum acceleration, or an obligation to visually signal autonomous driving. Overall, however, it must be stated that the operation of an autonomous vehicle without a responsible driver is not yet permitted in any country in the world.

4.2.3.2 Liability

For the question of liability in the event of an accident, it is generally decisive whether the cause of the damage was a responsible driver or the autonomous vehicle control system. For this reason, German road traffic law, for example, prescribes the installation of a black box for Level 3 and Level 4 vehicles, with the help of which it can be determined whether the vehicle was being driven by the driver or the system at the moment in question.

The question does not arise with fully automated Level 5 vehicles, where driver intervention is not envisaged and is technically no longer possible without a steering wheel or pedals. In this case, therefore, the legislator considers the manufacturer to be responsible in any case. In order not to have to wait for the passing of corresponding laws during vehicle development, manufacturers such as Volvo already voluntarily agreed in 2015 to assume full liability for damage caused by autonomous vehicles.

Civil liability and, if applicable, also criminal liability follows from the Product Liability Act, according to which the owner or operator is responsible for maintaining the vehicle in proper condition and thus has an operator responsibility. In contrast, the manufacturer must be held responsible for a failure of the vehicle control system that caused the damage. Here, too, in analogy to the applicable case law, it must be examined in each individual case whether there has been misconduct or negligence on the part of persons involved in the development and manufacture, which can then also lead to criminal law consequences.

In order to be able to react in the current test phase in the event of concrete damage, for example, the abovementioned law in California requires, as a prerequisite for the approval of driverless vehicles, the conclusion of liability insurance with coverage of at least five million dollars.

However, the long-term expectation of manufacturers and operators is that autonomous vehicles will drive completely accident-free and that the question of liability will thus also become obsolete.

4.2.4   Autonomous Flying

Private aircraft or private helicopters have always represented the epitome of exclusivity and luxury in the field of individual mobility. The high costs of purchasing and operating them, as well as of acquiring the necessary flight licenses, together with the strict legal requirements, mean that private aircraft remain the preserve of only an extremely small, wealthy customer group and thus hardly play a role in urban mobility systems. However, as soon as aircraft are controlled autonomously and can thus be operated without a pilot, this situation can change very quickly.

4.2.4.1   Autonomous Flying as Part of Mobility Systems

If the principle of the robo-cab, namely, the transport of individuals or small groups by small, driverless vehicles, is transferred to aircraft, significant potentials arise here. The concept of a *man-carrying drone*, i.e., an autonomous air taxi that can transport passengers including luggage within a conurbation, could be integrated into a wide variety of urban mobility concepts. The following reasons, among others, speak for this:

- The operation does not require any infrastructure apart from adequate take-off and landing facilities and possibilities for charging or refuelling. Air taxis are therefore ideal for routes that cannot be driven by car for topological or infrastructural reasons, such as crossing rivers, lakes, or mountains or overcoming large differences in altitude.
- For use in urban areas, small aircraft for one or a maximum of two passengers are required. The space, weight, and, of course, cost advantages resulting from the elimination of the pilot are particularly great in this size range.

- At the same time, aircraft of this size and with a range requirement of around 50 kilometers are ideally suited for the use of emission-free electric drives.
- The development of *multicopters*, i.e., helicopters with several rotors, is now so advanced that safe control is also possible in urban areas. Multicopters are *vertical take-off and landing (VTOL)* aircraft; i.e., they can take off and land vertically upward, which is a necessary prerequisite for operation in highly built-up metropolises, for example.

4.2.4.2 Realization

This potential has been recognized by the manufacturers of flying devices, the providers of mobility services (including the automobile manufacturers), as well as the municipalities, and they are working in various cooperation projects on the development of mobility offers with autonomous electric flying devices:

- Internationally, probably the most advanced in the development of an air taxi is the German start-up Volocopter in Bruchsal, which has developed a multicopter of the same name with 18 individually electrically driven rotors and space for two passengers. The Volocopter has been tested in Germany for several years (without "real" passengers) and has also been flying in Dubai since 2017, integrated into the airport's air traffic management system there. By 2028, Volocopters are to be integrated into the public transport systems of international metropolises, where they will then transport up to one hundred thousand passengers per hour.
- The US ride service provider Uber also presented a complete concept for the operation of autonomous air taxis early on with UberAIR. From 2020, the first test flights are to begin in Dallas, Los Angeles, and Dubai, with the launch of a broad, commercial offering of the services planned for 2023. Small, autonomous, electrically powered eVTOLs are to be used, which Uber does not intend to manufacture itself but rather to source from many different manufacturers. Cooperation partners here today are the Boeing subsidiary Aurora Flight Sciences,

Embraer, Bell Helicopter, Karem Aircraft, Pistrel Aircraft, Mooney, and for charging the company ChargePoint.

• Airbus is currently working on three concepts: The electric quattro-copter "CityAirbus" is expected to transport up to four passengers in 2023, although initially not autonomously but with a pilot on board. The "Vahana" tilt-wing aircraft with eight pivoting, electrically driven propellers is suitable as an autonomous air taxi for longer distances and is expected to be ready for series production as early as 2020, and, finally, the "Pop.Up" concept developed jointly with Italdesign and Audi, in which the passenger cell of a two-seater electric vehicle can be detached from the integrated chassis drive unit and docked onto an autonomous quattrocopter. The vehicle passengers thus become air-craft passengers without having to leave the cabin. In parallel, Airbus subsidiary Voom already offers on-demand helicopter flights in Mexico City and São Paulo and plans to integrate autonomous flights into its offering in the long term.

• Financed by the private fortune of Google co-founder Larry Page, the start-up Kitty Hawk is developing the autonomous two-seater "Cora" with 12 swivelling rotors. Kitty Hawk plans to operate these as air taxis in New Zealand.

• The German start-up Lilium Aviation is pursuing a similar concept. Their five-seater tilt-wing aircraft has a much greater range than other concepts and will be flown with a pilot for a transitional period. The considerable investment in Lilium Aviation (investors include LGT and Tencent) demonstrates the potential of this concept.

• Last but not least, the Chinese drone manufacturer EHang is working on the autonomous quattrocopter "EHang 184", which is to be used as a flying taxi. The development of the EHang is financially supported by the Chinese government, among others.

Even if it still sounds more like science fiction today and public opinion on the subject is still somewhere between sceptical and amused, autonomous air taxis are not only sensible but also technically feasible. In commercial aviation, 99 percent of all flights have been flown autonomously for decades. Today, every aircraft already has an *Automatic-Dependent-Surveillance-Broadcast (ADS-B)* system that transmits not only its own identification, position, and speed but also the planned flight route to air traffic control every second – in other words, ideal conditions for external control.

Teaching an autonomous multicopter to fly in airspace is much easier than teaching an autonomous car to drive in traffic. The reason for this is very simple: The number of possible situations that an aircraft can experience in the field – and which then have to be validated by tests during the testing phase – is much smaller than the ones that a car has to deal with in road traffic: In airspace, there are no children playing, no road works, and no black ice. It is therefore not only possible but to be expected that autonomous air taxis will be ready for approval even before the robo-cabs.

Even if it still sounds like science fiction: Flying autonomously is far less complex than driving autonomously. We'll probably be flying autonomous air taxis significantly sooner than we'll be letting a driverless taxi take us around town.

4.3   New Vehicle Concepts

4.3.1   Classification of Passenger Car Concepts

Over many vehicle generations, passenger car concepts have been differentiated according to two basic criteria: by *body type* and by *vehicle class*. The vehicle categories thus created, such as "mid-size SUV" or "luxury sedan", enable customers and manufacturers alike to classify and meaningfully compare market offerings. Analogous to sports, one knows immediately what is involved: the body type corresponds to the sport being played and the vehicle class to the league being played in.

Which body type a customer chooses, for example, a coupé, a convertible, or an SUV, depends on their personal necessities and individual preferences. Specific requirements in terms of space due to job, hobby, or number of family members can be restrictive or decisive as hard criteria in the choice. Another decisive factor is often the public image of the vehicle type in question: Do I, as the owner and driver, want to be perceived as sporty, dominant, or elegant? Should the car stand for presence and sophistication or for simplicity and understatement? Macro- and micro-societal trends play a major role here, such as a possible general disapproval of SUVs among friends.

Over time, vehicle length has established itself as a quasi-universal, quantitative indicator of vehicle *class*. Common model designation sequences of manufacturers, such as 1 Series, 3 Series, 5 Series, and 7 Series; C-Class, E-Class, and S-Class; or A2, A4, A6, and A8, correspond to certain increments of vehicle length, with which comfort, equipment level, engine performance, safety, and ultimately the price also gradually increase.

4.3.2   New Criteria for Classification

However, current mobility trends are breaking down the established, one-dimensional relationships in vehicle classification. The demand for electric drives and mobility services, the shortage of parking space, and the increasing and stricter legal regulations require and enable new vehicle concepts as well as model and brand strategies of the manufacturers.

- The use of electric drives enables completely new vehicle concepts: The engine and control unit require significantly less space than an internal combustion engine. The gearbox and exhaust system, including thermal management, can be dispensed with, but the battery, a relatively large and heavy additional component, must be accommodated in the vehicle, ideally in the vehicle floor. In the end, there is significantly more space in the electric vehicle. For example, the external dimensions of a BMW i3 are roughly the same as those of a BMW X1, but its interior is comparable in size to that of a BMW X3. The possibility of distributing the required drive power among several wheel hub motors that can be integrated into the wheels creates additional space potential.
- Vehicles that are primarily designed to meet the requirements of passengers will look completely different from vehicles that are classically designed only for the demands of the driver. With passenger-oriented design, the focus will be on the interior, comfort, and connectivity, while at the same time the demand for engine performance, agility,

and dynamics will drop significantly. Incidentally, this applies completely irrespective of whether the vehicle is controlled by a driver or autonomously.

- Especially in urban areas, the size of a vehicle is often perceived as a nuisance and disadvantage, not only when looking for a parking space. The – theoretically available – power of a powerful engine can also only be called up here very rarely but at the same time leads to ever higher operating costs or even emission-related access restrictions. Conversely, the agility of smaller vehicles brings clear advantages in urban traffic.

No wonder, then, that manufacturers are experimenting with new concepts that combine high ride comfort at urban speeds and a comfortable level of equipment with small vehicle dimensions. The task of vehicle development here is to decouple vehicle and brand characteristics such as status, presence, or sportiness from the vehicle dimension and to transfer them to a new, urban context. A corresponding thought experiment here would be what a 3-meter-long city car from Rolls-Royce, Bentley, or Bugatti would look like.

An early and at the same time extreme approach in this sense is represented by the so-called *Kei Cars* in Japan, a vehicle class with strong tax benefits and a restrictive limitation of the maximum vehicle dimensions to a length of 3.4 meters, a width of 1.48 meters, and a height of 2 meters, as well as a maximum engine capacity of 660 cubic centimeters and a maximum engine output of 47 kilowatts. In this segment, a number of manufacturers have impressively demonstrated how particularly comfortable, sporty, or even luxurious vehicles can be realized even within such a narrow regulatory framework.

Car users' priorities regarding vehicle features and equipment are beginning to shift.

For future offerings, the traditional categorization by vehicle type, vehicle size and engine power will no longer be sufficient.

4.4   Digitalization

Electromobility and autonomous driving, the trends discussed in the previous two chapters, are ultimately innovative vehicle functions that represent alternatives to today's standard solutions of *drive by combustion engine* and *vehicle control by driver*. The question of whether and when these alternatives will one day completely replace the existing solutions is one that no one can answer with any degree of certainty today – as can be clearly seen from the strongly divergent statements of the available forecasts. It is much more likely that electric motors will replace conventional drives to a significant extent than that only autonomous vehicles will be driving on all the world's roads; nevertheless, individual mobility will probably continue to involve not only steering wheels but also internal combustion engines for the foreseeable future.

In contrast to these rather slow and quasi-continuous processes of change, *digitalization* is having a much faster, more comprehensive, and thus more dramatic impact in many business areas. On the one hand, vehicles are expanding their functionality through digital services, but on the other hand they themselves are becoming part of new digital mobility services and are then no longer at the top of the value creation pyramid but on the second level. Also, this is precisely where the disruption potential of digitalization lies, in the possibility of quickly and completely replacing established business models that have been successful for a long time. Today, vehicle manufacturers and providers of mobility services are confronting the risk of oversleeping like Kodak in digital photography or like Quelle in online retailing in very different ways. In order not to miss the digital train and to be able to play along with the new offers, most of them are therefore deliberately looking for cooperation with digital players. In this global search for partners, however, not only do very different technologies and business processes come together but above all dramatically different cultures.

4.4.1 The Five Stages of Digitalization

Hardly any other term has dominated the technology and business media in recent years as much as *digitalization* – and hardly any other term is interpreted and used so differently. However, a reasonable assessment of the performance and potential of digital services requires a differentiation and clarification of the term. If we trace the history of digitalization, we can distinguish between five successive stages of development, each of which has enabled or will enable certain products, functions, or services and only in combination lead to the disruptive potential mentioned above: digitalization as a move away from analogue data, connectivity, mobile Internet access, the spread of smartphones, the targeted analysis of large data sets known as Big Data, and machine learning from this data.

4.4.1.1   Stage 1: Digitalization in the Literal Sense

Literally, digitalization means the conversion of analog information, such as a printed text or a photo, into digital data that can then be stored, duplicated, processed, and sent without loss. With the spread of personal computers in the 1980s, this first stage of digitalization not only created new fields of business but also saw the disappearance of previously very successful companies whose products or services were now no longer needed. Digital photos no longer require optical cameras, films, film processing machines, or service providers. Digital text documents require neither typewriters nor photocopiers, neither file folders nor the associated shelving. Digital drawings require neither drawing boards nor drawing pens and neither rulers nor paper and duplicating services. These examples clearly show how even the first and simplest form of digitalization is already permanently calling into question entire branches of industry.

However, the impact on mobility is limited at this stage. Printed road maps, such as the formerly ubiquitous Shell atlas, have been almost completely replaced by navigation devices with digital maps, and mechanical engine controls have given way to digital ones. The main impact of the first stage, however, is not on customer offerings themselves but on manufacturers' internal business processes – such as the use of CAD systems in vehicle design instead of pen and drawing board.

4.4.1.2   Stage 2: Connectivity

The interconnection of previously isolated, local computers via the Internet to form the *World Wide Web* then enables the second step of digitalization, namely, the rapid and worldwide exchange of digitally available information, and thus replaces existing analog modes of communication: e-mail and online ordering replace letters and telephone calls, the possibility of searching for information worldwide via search engines replaces costly research services, and local databases can be linked with each other and thus automatically compared. The technical prerequisite for connectivity is, in addition to the realization of the physical and logical data connection (such as the deep-sea cable for the intercontinental connection), the agreement of standardized exchange formats for the data to be transmitted.

Connectivity has also created business models that are now indispensable and have replaced those that existed until now. For example, it enables local searching, ordering, and payment for products and services and thus forms the basis of online retailing, which has already replaced large parts of traditional retailing. Another area in which processes have changed radically as a result of connectivity is internal collaboration between the usually worldwide distributed sales, production, and development sites of industrial companies.

In the area of mobility, connectivity primarily offers advantages to users of mobility services: Current timetables and price lists for local and long-distance public transport can be called up directly, and services can be reserved and paid for online. Vehicle buyers can also call up current information on relevant models and visualize, configure, order, and pay for vehicles on the screen. When it comes to using the vehicle itself, on the other hand, the effect of connectivity is still relatively limited. Since the vehicles themselves are not yet connected to the Internet, pure connectivity does not yet lead to any increase in functionality. Users can, for example, access digital map updates or music files on the Internet from their local computer, but transmission from the computer to the vehicle then takes place via cable or USB stick.

4.4.1.3 Stage 3: Mobile Access to the Internet

Since the beginning of the 2000s, mobile technologies such as UMTS and later LTE have enabled nationwide, fast, and inexpensive Internet access beyond the site-bound LAN networks. This means that non-local computers such as laptops and handheld devices, as well as vehicles of all kinds, can be connected to the Internet and communicate with each other. Wireless mobile network access also enables worldwide access to clouds, i.e., distributed synchronized servers, in order to use their storage space, computing power, or application software. Direct network access in the vehicle is usually implemented via a mobile phone brought in by the user or a vehicle's own SIM card.

Wireless mobile connectivity enables vehicles to exchange data with stationary computers (*backends*) and with other vehicles in near real time. Data transmission to the vehicle (*downstream*) enables new customer functions such as current news services, automatic map updates, traffic flow information in real time, or music downloads; data transmission from the vehicle to the backend (*upstream*) enables, for example, the analysis of driving style, the health status of the vehicle, but also the collection of environmental data collected by the vehicle, such as traffic signs.

As a result, the permanent connectivity of vehicles represents the basis for all data-based services and business models in the field of mobility. In particular, vehicles are becoming active components of the *Internet of Things (IoT)*. This enables interaction, for example, with connected devices or connected infrastructure at home – for example, by automatically opening the shutters and raising the room temperature when the vehicle approaches the house.

4.4.1.4   Stage 4: Smartphones

Worldwide, fast, and inexpensive network access is then in turn the basis for the worldwide spread of smartphones. In addition to their primary use, namely, being able to access the Internet at any time and anywhere in the world with the installed apps, the sensor technology integrated in them in particular opens up completely new possibilities: The cameras integrated into smartphones have caused sales of conventional digital cameras to plummet dramatically, access to huge music and video libraries has presumably dealt the death blow to CDs and DVDs, and the continuous recording of one's own position via GPS enables *local based services (LBS)*, i.e., information and service offers tailored to one's current location and, if necessary, also one's personal behavior.

This option in particular opens up completely new possibilities in the field of mobility: Services such as car sharing or ride hailing can only be meaningfully presented as LBS. If you can see on your smartphone where the next taxi or the next subway station actually is, you will be much more willing to use these services and to forego driving your own car. The fact that the smartphone becomes a navigation system simply and at no extra cost by downloading digital maps and tracking one's own position has put their providers under considerable pressure. At the same time, the connection between smartphone and vehicle enables a host of new functions, such as remote control of vehicle locking or auxiliary heating via app – or information relevant to the driver, such as weather or road conditions, which relate to the vehicle location or the planned route. Smartphone access to vehicle data via corresponding *application programming interfaces (API)* and the mobile Internet also allows extensive diagnostic functions.

4.4.1.5   Stage 5: Big Data, Analytics, and Artificial Intelligence (AI)

A growing number of computers, smartphones, connected vehicles, and other things around the world make a huge amount of data available online, today commonly referred to as *Big Data*. With the help of appropriate *analytics tools,* this data can be analyzed in a targeted manner and further processed into insightful information: Which functions of a vehicle are used most often – and which are not used at all? Which contents of a website are attractive – and which are not? What leads to the abandonment of an online purchase process? This information can be used to derive customer- and requirement-oriented offers, for example, in the sense of "customers who bought product X also bought products Y and Z".

For example, Big Data technologies can also be used to record and evaluate mobility behavior within a mobility area as a whole. The analysis of the available movement data from smartphones and vehicles allows a comprehensive picture of traffic flows, shows the course of the usage rates of different means of transport, and allows a wide variety of evaluations, such as "How strongly does the degree of usage of car sharing in Berlin depend on the weather?"

4.4.2   Digitalization Among Vehicle Manufacturers

Whether bicycle, car, bus, or train, before a vehicle can be used, it must be developed, produced, and sold. After it has been taken over by the customer, it must be regularly maintained and repaired if necessary, and questions from the customer about functions or problems must also be answered. All of these activities are the responsibility of the *original equipment manufacturer (OEM)* – which makes it clear that although this term has become commonplace, it does not adequately describe the scope of tasks of the companies it refers to. The added value of a manufacturer encompasses much more than just production, namely, also development, distribution, and aftersales.

So when we talk about digitalization at vehicle manufacturers here, we are talking about the extensive effects and potential of digital technologies and methods in all relevant processes – in development, purchasing, production, marketing, sales, customer care, and aftersales. Internal business processes are becoming faster, safer, more cost-effective, and more flexible thanks to digital methods such as connectivity, mobile data collection, and analytics. This is mostly demonstrated here using the example of processes of passenger car manufacturers, as these are the most extensive and complex compared to those of other vehicle manufacturers.

It is striking that, as described in the next chapter, the vehicles themselves already have extensive digital functions, but their manufacturers are obviously finding it much more difficult to use the possibilities of digitalization in their internal processes as well. There is still a great deal of potential to be tapped here.

4.4.2.1 Digitalization in Vehicle Development

Requirements Clarification

What does the vehicle that the manufacturer wants to bring to market really has to be able to do? Which features and functions does the customer really need – and which are less important? In the conventional approach, customer behavior is determined via customer surveys in order to clarify the requirements for vehicles or services. The results are then used to estimate usage profiles, such as how dynamically sports cars are actually driven. However, such survey results are subject to strong distortions, on the one hand, due to the random sampling (perhaps only sporty drivers take part in such a customer survey) and, on the other hand, due to the subjective assessment of one's own behavior (hardly any sports car customers will state that they drive rather slowly and leisurely – even if they do). Also, some questions the car owner may not even be able to answer, such as how often the right rear window is actually opened and closed again (which is certainly relevant for the design of the window drive and thus for its costs). Another disadvantage of surveys is the long time it takes for the results to become available: The development team ultimately has to wait until the market research agency has finally produced reliable statements.

By contrast, the online collection of data from vehicles in the field allows usage behavior to be recorded and evaluated quickly, comprehensively, and objectively. For manufacturers, this analysis yields valuable and often surprising insights. For example, the fact that customers also

use seat heating intensively in summer – most likely because of its benefi-
cial effect on back pain – must be taken into account in its design. The
manufacturing costs for footwell lighting that no customer has ever
switched on can certainly be saved. By linking the usage data of different
functions, additional insights can then be gained regarding customer
behavior, such as the temporal relationship between the setting of the air
conditioning, the opening and closing of windows or sunroof, and the
outside temperature.

In addition to the usage data collected in the vehicles, additional data
sources such as social media analyses or the evaluation of customer service
calls are available when clarifying requirements. These are collected continuously, resulting in huge amounts of data over time. A targeted evaluation of this data that makes sense in terms of effort and results is possible
with the help of big data technologies such as data warehousing or business analytics.

*The possibility of continuously recording and centrally evaluating data relating
to vehicle use across the entire fleet leads to quantum leaps in the quality and
speed of information when clarifying customer requirements in product planning and vehicle development.*

Product Design

In vehicle development, the clarification of requirements is followed by
design – an area in which digital working methods have been used for
decades. Experts from design, engineering, and testing as well as production planning work together on computer-internal, digital product models using IT system-supported methods such as *computer-aided design
(CAD), computer-aided engineering (CAE)*, or *computer-aided planning
(CAP)*. These can be virtual vehicles, but also, for example, virtual factories or workshops. Virtual vehicles combine the exact geometric shape of
the product and its components with technological data such as material
properties as well as organizational and commercial data such as prices or
component numbers; virtual factories represent machines, plants, and
logistics processes.

With such digital models, products and processes cannot only be developed quickly and efficiently, but they are also a fundamental prerequisite for *cooperative development*, which is essential in many areas today, in which experts in different locations work together on a development task at different times and can always access the current data status. *Virtual reality (VR)* or *augmented reality (AR)* techniques are used for the realistic visualization of and interaction with digital product models. In this way, both the vehicle and the associated processes and services can be "experienced" even without test hardware and thus very early in the development process and optimized by evaluating and selecting alternative solutions.

Even if virtual vehicles are already old hat in the automotive industry, the potential that lies dormant in their use is far from exhausted. The goal should be that all the knowledge that can be gained from virtual vehicles is also gained through analysis, evaluation, calculation, and simulation. A prototype based on such a mature design status will then only yield a few new findings during testing and thus require few then time-consuming and costly modifications. Today, the virtual car process is far from this ideal. Many developers prefer to wait until the first prototypes are available and accept that many test results are obsolete because they are based on outdated designs. The emergence of a generation of developers who are more open to agile working with digital product models can still exploit extensive potential here.

In addition to the use of digital product models, digitalization has recently brought benefits to development departments on a completely different level. Applying agile principles developed for the creation of software, such as development in fast cycles (so-called *sprints*) or the early realization of a minimum solution for the customer, the so-called *minimum viable product (MVP)*, to the development of vehicles and services, can result in significant time and cost advantages. The problem is that agile principles are often in marked contrast to established ways of thinking and acting in companies (cf. Sect. 4.4.6). Their successful introduction requires nothing less than a change in established working and management models and represents a real challenge for management.

4.4.2.2 Digitalization in Production

As in vehicle development, digital methods and techniques also made their way into vehicle production at an early stage. In the 1980s, companies introduced *computer-integrated manufacturing (CIM)*, the interconnection of production facilities, and the systems for development, planning, and accounting. From the digital description of the product and the planned production process, for example, the control programs for the corresponding production facilities were thus derived directly. However, the ideal of the CIM idea of interconnecting all systems and processes involved in value creation, and providing them with a common, consistent database has only been implemented in niches, if at all. The reasons for this are, on the one hand, the complexity of the resulting overall system and, on the other hand, simply the many existing (and paid for!) but not connectable production facilities in ongoing operation, which cannot simply be switched off and sold.

Under the impression of the opportunities and risks of the "new" digitalization, the idea of complete connectivity in production has been taken up again worldwide in recent years, for example, in Germany through initiatives called *Industry 4.0* (which is intended to refer to the forthcoming fourth industrial revolution through digitalization) or in the USA through the *Industrial Internet Consortium (IIC)*. However, while the virtual product and the virtual factory were rather static models, the idea of a *digital twin* developed today in the context of Industrie 4.0 represents a digital real-time image that can be evaluated and analyzed at any point in time via analytics and AI techniques, allowing, for example, errors to be detected or knowledge-based predictions to be made and thus processes to be optimized. The possibility of wireless connectivity also makes it possible to include the logistics chains for individual parts and finished products.

4.4.2.3 Digitalization in Marketing and Sales

Online Product Configuration

Before a potential vehicle customer makes a purchase decision, he not only wants to obtain detailed information about the vehicle he is interested in, but he also wants to compare options and experience it as far as possible. In this phase, manufacturers' websites are increasingly serving as virtual showrooms in addition to visits to the dealership. More and more customers, especially young customers, prefer to inform themselves at home via the Internet rather than arrange a consultation at the dealer's premises, where they feel they are being put under pressure. Wherever vehicles that are not fully configured are sold directly from the dealer's yard or only models with a few equipment alternatives are available, online configurators play an important role here, allowing customers to select their model; select and visualize the available colors, engines, upholstery, or special equipment; and compare them with other alternatives. The individual virtual vehicle created in this way can then be viewed on the screen in all its details, and the price for it is also always transparent.

For the success of marketing and sales, the quality of this Internet presence can hardly be given enough importance. An inspiring, individual user experience of the Internet presence considerably increases the chance that a vehicle purchase will be made at the end of the sales funnel. If, on the other hand, you have to slog through confusing menus, wait for long periods of time before the image is displayed and endure error messages again and again; if you are constantly confronted with offers that do not correspond to your own preferences and needs, you will quickly abandon the process and may be lost as a customer. In this sense, conducive design principles for online configurators are – in addition to reliability, speed, and timeliness – for example, *gamification* and *individualization*:

* *Gamification* is the application of elements known from computer games to create an entertaining and inspiring user experience. The goal is the viral spread of the website far beyond the direct target group. If the configurator of a brand is really fun, if you can configure your

dream car there, save it, print it out, or share it with friends via social media, maybe even export it and use it in racing games; you also bind people to the brand who actually didn't want to buy a car (yet).

• The basis for *individualizing* the offer is, on the one hand, the master data and preferences stored by the user in his personal account and, on the other hand, the evaluation of customer behavior when using the configurator through analytics tools. For example, which model is suitable based on purchase history, driving behavior, or family situation? Which one should be recommended first? For example, someone who frequently researches sports cars is probably more interested in dynamics than comfort when selecting options, even if they are currently configuring a station wagon suitable for families. Analogous to Amazon's "Customers who bought X also bought Y", additional equipment and packages can also be recommended here. Also, maybe there are options that the prospective buyer has looked at more often and thus considered more intensively but then didn't choose after all? In this case, it would probably be helpful to provide further targeted information that highlights the advantages of this option even more clearly.

Online Sales

More and more customers want to order and pay for or finance the vehicle they have configured online. The young, up-and-coming generation of buyers in particular has grown up with online purchases and has no problem whatsoever with purchasing not only clothing and household appliances but also making purchases on the scale of a car on their computer or smartphone. More and more manufacturers are offering this option as well. An A.T. Kearney study from 2016 predicts that, by 2020, more than a third of all new vehicles worldwide will already be sold online.

While selling cars online is practical and convenient from the customer's point of view, it represents a significant business risk for the largely independent dealers. This is why it must be regulated in the contractual agreements between dealer and manufacturer whether the latter leaves the online sales channel to its dealers. The fundamental possibility for the manufacturer to sell vehicles online directly to the customer, bypassing the dealer organization, weakens the negotiating position of the dealers in any case. If the manufacturer were to take over online sales itself, the dealers would still be left with test drives, delivery, maintenance, and repair as part of their business. The manufacturers, in turn, depend on the dealers to provide these services, which is why the cooperation will always be agreed in the end in a form that is profitable for both partners.

However, a further risk – for manufacturers and dealers – lies in the "hostile" takeover of online sales by third parties. Cross-manufacturer Internet portals, where technology, appearance, and prices can be compared across all manufacturers and brands and vehicles can be purchased, would squeeze manufacturers and dealers out of the customer interface. Even here, however, the aforementioned business content remains necessary, requiring physical interaction with the vehicle. After all, even those who buy a car online may want to take a test drive and definitely want to know which workshop they can contact later if necessary.

Post Contract Marketing

Once the customer has ordered the vehicle, digital methods can then be used to exploit the sometimes relatively long time until delivery of the vehicle in terms of customer loyalty and to generate further contribution margins. For example, customers can be enabled to track their vehicle step by step during this process. Analogous to the status messages for orders in online retail, the vehicle customer can also be regularly informed about the status of production and delivery, for example, that the assembly of the vehicle has begun, the engine has been assembled, or the vehicle has left the factory. The customer's anticipation can be taken into account here by including photos or video sequences. In addition, the customer can also be offered a *Next Best Offer (NBO)* at this stage: Depending on the season, those who have ordered a vehicle with all-wheel drive may now be receptive to a ski rack or winter tyres, while those who have opted for a high-powered variant may be receptive to driver training.

Additional business potential lies in the temporary activation of vehicle functions during the usage phase, which is made possible by the connectivity of the vehicles. Tesla customers, for example, can already have autopilot activated after purchase. When using mobility services, for example, this makes it possible to book additional features such as music streaming or a sunroof for an additional charge. On the one hand, this requires the availability of the necessary hardware in the vehicle, but, on the other hand, it also requires the cooperation of the specialist departments from sales, aftersales, and customer service.

The ability to switch vehicle functions on and off "over-the-air" can be translated by manufacturers into completely new business models. However, these break down the classic boundaries between sales, aftersales and customer service.

4.4.2.4 Digitalization During the Use Phase

While digitalization in development, production, and sales is primarily aimed at optimizing internal business processes, the focus is on something else in the usage phase after the vehicle is handed over: the collection and evaluation of customer and vehicle data for the realization of data-based vehicle functions and for customer service.

Data-Based Vehicle Functions

Conventional software-controlled functions in the vehicle, such as the CD player, door locking, or ABS, are self-contained, mechatronic systems. The necessary hardware components are assembled in the production plant, and the associated control software is loaded onto the respective control units. The function is thus available at the end of the production process, its quality is determined and can be checked and verified when the vehicle is handed over to the customer. As long as no errors occur, the function is then permanently available without further external input. Navigation systems represent a borderline case here, for whose functionality external data is not necessarily required, but the GPS signal received from outside is.

Digital services and functions, on the other hand, require the exchange of data with the vehicle environment. This is implemented via a mobile Internet connection, with a *backend server* operated by the vehicle manufacturer usually interposed between the vehicle and external servers for security reasons. This permanent connection of the vehicle with the Internet, in interaction with smartphones and external data providers, enables a variety of new functions in different categories. Examples of this are the following:

- Convenience:
  Operating the auxiliary heating, locking the doors, locating the vehicle's position, or retrieving information on the vehicle's condition via smartphone. Automatic closing of the convertible top when it rains …
- Infotainment:
  Individualized access to breaking news, streaming music, and video both directly through the vehicle and via personal smart devices brought in the vehicle …
- Journey Management:
  Transfer of destinations from own devices into the navigation system. Calculation of the route in the cloud. Online update of digital road maps. Traffic flow info. Localization and reservation of charging stations …
- Safety:
  Warning of obstacles caused by vehicles in front; automatic emergency call in the event of an accident with forwarding of relevant data on vehicle position, number of occupants, and the course of the accident (e-call) …

In order to be able to guarantee the function of such digital services in the long term, the operational readiness of the vehicle's system components alone is not sufficient. Rather, the vehicle must exchange data with its environment at different frequencies. For the traffic flow information function, for example, it must transmit its current position in order to receive the appropriate congestion information back from the data provider, if possible every minute, which is then visualized in the navigation system. For this to work reliably, the mobile Internet access, the backend, and, if necessary, other elements of the data transmission must be permanently available.

To ensure the operational readiness of digital services, vehicle manufacturers must therefore make a permanent effort, in contrast to conventional software-controlled functions: The quality and deliverability of the required data must be ensured at the supplier, and the backend and mobile Internet connection must function and be kept stable. For this reason, digital services are also offered in other business models: not as one-off special equipment that is reflected in the purchase price of the vehicle but as services that can be booked for a certain period of time. However, it has been shown that the implementation of the processes for booking, payment, and activation or deactivation of these vehicle functions often represents significantly more effort than the technical implementation of the actual vehicle function.

Digital Services in Aftersales

In the perception of customers and companies, aftersales is often overshadowed by vehicle sales. To put it somewhat pointedly: anyone intending to buy a new car sits at the dealership in attractive lounges with designer furniture and is welcomed and advised there by well-dressed sales staff with cappuccino or prosecco; anyone coming later for an oil change, to have winter tyres fitted or to buy a roof rack, on the other hand, stands in the queue in the open-plan office opposite the parts store and waits for a friendly "Next please!".

However, that is about to change. For hardly any other area is the potential of digitalization as high as for aftersales. The following points show just a few examples of how the connectivity of vehicles can contribute to business management, the perception of innovation, and customer satisfaction in aftersales – always assuming, of course, that the customer consents to the use of data:

- Online booking of service appointments:
  Connection with the IT systems of the service companies allows the vehicle user to conveniently make maintenance or repair appoint-

ments online instead of by phone – from the car, by smartphone or on the computer. The relevant data is transferred directly from the vehicle so that the service company can optimally estimate the time and spare parts requirements.

- Early detection of vehicle problems:
  Whether engine oil temperature, the voltage level of the starter battery, error entries in the transmission control unit, or noises recorded via the microphone of the hands-free system – networked vehicles can generate a wide range of data via the sensors and control units installed in them, with the help of which technical problems cannot only be detected, but also diagnosed or even predicted using artificial intelligence. As a result, broken down vehicles can be avoided or repairs can be carried out at an early stage, thus reducing consequential damage. Through the early detection of spare part requirements for the repair, these can then be ordered in advance, thus further reducing the duration of the workshop visit.

- Determining the health status of vehicles in the field:
  If the same problem occurs in a large number of vehicles, for example, due to a systematic error in vehicle assembly or in the manufacture of supplier parts, the manufacturer usually finds this out today on the basis of the warranty costs claimed from the dealers. The connectivity of vehicles offers the potential to detect and remedy such systematic errors up to several weeks earlier, which means that fewer defective vehicles are delivered that have to be repaired later as warranty cases. In the case of design-related faults, this procedure also enables the rapid integration of design changes into the ongoing production of the model.

- Remote software update:
  By updating the vehicle software, functions can be improved or problems solved. Today, customers usually have to drive to the dealer, where the vehicle is connected by cable to a corresponding programming system and left there for several hours. With networked vehicles and a correspondingly powerful mobile Internet connection, however, such an update is also possible "over-the-air". As with smartphones, Tesla, for example, already offers its customers regular fixes and functional enhancements in this way. However, it must be ensured that a func-

tioning software set is always stored in the vehicle as a backup. If an error should occur during the transfer of the new version, this can be used to restore operability even without an external data connection.

- In-car sales:
Online offers for accessories and services represent another option for digital services in the vehicle. From tips on nearby restaurants or cafés after several hours of driving to the option of streaming music and videos for a fee to the offer of winter tires at the beginning of the cold season, additional sales can be generated via online sales in the vehicle across a broad spectrum of offers. However, the actual implementation must be carried out with a high degree of sensitivity for the vehicle customers; the threshold beyond which such offers are no longer perceived as support, but even possibly as a nuisance, is quickly reached.

- Customer care:
In addition to the telephone hotline, where more or less competent employees answer customer questions with the help of expert systems, digitalization also offers significantly more opportunities here to provide support in solving product problems or to answer questions in a more informed manner. As is the case with online sales or other digital services, customers can contact specialists via forums and chat functions, who can also connect to the vehicle and connected systems to analyze the problem, use analytics tools, and also resolve certain problems "over-the-air".

Change from Manufacturer to Service Provider

What car manufacturers in particular have to learn is that this also expands the focus of their business model from the above-described sale of the best possible product to the customer – to include the permanent operation of data-based vehicle functions and the provision of supplementary services throughout the entire usage phase of the vehicles produced. They are moving from being a manufacturer to being a service provider.

A factory-fitted navigation system, for example, accesses the digital map and the GPS signal received in the vehicle in order to calculate the optimum route to the destination using the software that is also on board. The manufacturer does not have to do anything further for this function. As with the drive, the window regulator, or the vehicle lighting, the following applies: As long as no defect occurs, the system works.

In contrast, the integration of real-time traffic flow information into the navigation system is perceived by the customer as a mere functional extension of this system (the route is now shown in green, yellow, or red on the display according to the traffic density), but it represents a completely new situation for the manufacturer: To ensure that the function runs reliably at all times and that the driver is shown the current traffic situation, the manufacturer must now ensure around the clock that his data provider sends him correct and up-to-date traffic flow data to the backend, that the backend takes over the intermediate storage and distribution of this data, and that the mobile Internet connection is up and running via which the individual vehicles send their request to the backend and then receive the corresponding traffic flow data from there. The situation becomes even more dramatic with the control of autonomous vehicles, the most complex data-based vehicle function.

In the same way that Apple or Samsung regularly update the operating systems of their smartphones after the sale, provide services in the form of apps, and operate some of these themselves, car manufacturers are thus also suddenly turning from pure product sellers into function operators and thus into service providers. They are now in permanent contact with customers far beyond the previous scope of aftersales, know their usage behavior and the condition of their vehicle, and provide them with data-based services and offers – for as long as the vehicle is in operation, i.e., even when it has long since reached the third or fourth owner. At the same time, the perception of quality and brand is no longer defined solely by the product substance of the vehicles but increasingly also by the experience of the associated services. Not only a vehicle can be innovative, but also the way in which a workshop visit is organized, from making an appointment to returning the vehicle, can contribute to the corresponding positive or negative perception.

On the one hand, this often underestimated change opens up access to completely new, attractive business models, but, on the other hand, it also requires nothing less than a complete rethinking of the product concept, the customer interface, and the understanding of quality. At the same time, however, many manufacturers have not even clarified which division is responsible to ensure the long-term operation of the data-based functions. Whether traffic flow display, music streaming, or e-call, the responsibility for such services is neither properly located in development nor in production nor in sales. However, since the interaction with the "driving customer" (as opposed to the "buying customer") is managed by aftersales, the operator responsibility for data-based services is basically also in good hands there. However, the structures and processes of aftersales at most manufacturers today are designed for maintenance, repair, and spare parts sales and not for permanently ensuring the functionality of digital services in and around the vehicle.

The connectivity of vehicles makes it possible to implement attractive new vehicle and service functions. In the future, it will be up to the after-sales departments of the manufacturers to operate these permanently and reliably in the field, and for them this is both a growth opportunity and a technical and organizational challenge.

4.4.2.5 Data-Based Services for Other Mobility Stakeholders

The digital services during the use phase discussed in the previous chapter are primarily tailored to the owner or driver of a vehicle. However, the connectivity of vehicles can also provide useful and valuable services for other stakeholders in individual mobility, especially for passengers, but also for passersby in the immediate vicinity of the vehicle or even the municipality in which a vehicle is currently moving.

Digital Services for Passengers

The ability to connect vehicles permanently or temporarily to the Internet via a backend enables a range of services for all vehicle occupants, not just family members or colleagues but especially customers of ride-sharing services such as taxis or Uber. Examples of such services are:

- Providing an Internet connection via the vehicle (hotspot)
- Possibility to adopt stored, preferred personal settings in the vehicle, for example, for air conditioning, seat, or radio stations
- Pairing the vehicle with embedded smartphones or tablets to display information or entertainment via vehicle displays
- Display of information about the journey such as route history, POIs along the route, and estimated time of arrival, if necessary also with translation into your own language
- Possibility to earn "miles" as a customer of rideshare services for booking services with vehicles of a certain brand within the framework of a brand-related loyalty program

Such functions would then have to be offered via corresponding interfaces in all vehicles of a manufacturer or even across manufacturers; operation would then not take place via the respective vehicle, but via a passenger app of the vehicle manufacturer on the passenger's smartphone.

Such a range of passenger-focused services offers vehicle manufacturers the opportunity to attract and retain mobility customers who do not own a vehicle of their brand. Those who can access the abovementioned comfort functions in vehicles of a certain brand via a passenger app, for example, will increasingly book mobility services in which vehicles of this very brand are used.

Digital Services for the Community

The fact that, particularly in metropolitan areas, more and more people, and not only younger people, are foregoing owning a car and using mobility alternatives instead has already been mentioned. This is accompanied by the demand of these citizens, formulated in their capacity as voters, to the respective municipal administration to reduce the number of private motor vehicles in the city in order to reduce emissions and to reclaim public spaces previously reserved for driving and parking motor vehicles for other types of use. This trend, which is obviously more than worrying for the traditional business model of vehicle manufacturers, will be discussed in detail in Chap. 6.

However, what if a vehicle benefits not only the driver and passengers who drive or are getting driven by it but also, for example, passersby in the vicinity of the vehicle or, more generally, citizens of the city through which it passes? The following ideas illustrate what effective data-based functions could look like in this direction:

- Vehicle as a rescue cell:
  Vehicles can be used as a rescue cell by passersby in an emergency situation. By pressing a corresponding emergency button attached to the vehicle, it can be opened and then closed again from the inside, thus offering protection from possible attackers. At the same time, it triggers an emergency call, records the situation around the car with the vehicle cameras and alerts other passersby via the vehicle lights and acoustic warning signals.
- Vehicle to assist in emergencies:
  If necessary, the boot or another compartment of the vehicle can be opened from the outside and the first aid kit stored inside can be used. At the same time, a telephone connection to the rescue control center can be established via the vehicle's Internet connection.
- Vehicle as a pedestrian warning:
  Vehicles parked at the side of the road use their exterior cameras to detect pedestrians (especially children) stepping onto the road between them and warn passing traffic via the vehicle lights.
- Vehicle as info counter:
  Vehicles are online and have a touch display accessible from the outside, which passersby can use to call up public transport connections or current local information, for example.
- Vehicle as a wifi hotspot:
  Vehicles are online; a sign informs passersby that they can be used as Internet hotspots.

Even if some of these functions obviously lend themselves more to vehicles from fleets of mobility service providers than to private vehicles, cars whose presence ultimately benefits the urban community would certainly end up being the last to disappear from the city, even from the perspective of car-averse citizens.

Digital Services for the Benefit of Municipalities

As far as road traffic is concerned, traffic flow, parking space, and traffic safety are the main problems for cities today, in addition to vehicle emissions. No wonder, then, that local governments are discussing a wide variety of ways in which the permanent connectivity of vehicles, infrastructure, and other "things" can be used in the IoT and digital methods can help to solve these problems. By their very nature, the approaches being considered are not always in the interest of the individual motorist – as the following two examples show:

- Central traffic management:
  One approach already mentioned in Sect. 3.2.3, which is being studied in particular in megacities such as London, is the implementation of a central, urban traffic management system. Just as the control system in the distribution center of a parcel service ensures that hundreds of thousands of parcels delivered by trucks every day are moved as quickly as possible to the warehouse and from there on to the delivering vans through optimal use of the logistics infrastructure, a central traffic control system should optimally distribute all vehicles operating in a city to roads and parking spaces and thus bring traffic flow and parking space availability to an overall optimum. To do this, each vehicle that wants to leave or enter the city limits must first send its current location and destination to the traffic management system. The system then uses big data techniques to determine a route and parking space for each vehicle, so that an optimal utilization is achieved across the entire city area. The route and parking space can, of course, deviate from the driver's wishes, but they are fed back to him or her as mandatory specifications. Any deviation from these specifications is immediately detected by the traffic management system and taken into account or sanctioned accordingly.
- Integrated traffic monitoring:
  An approach that is much easier to implement technically, but which encroaches even further on the personal freedom of action of vehicle drivers and owners, is the implementation of traffic monitoring inte-

grated into the vehicle. Today, every connected vehicle already "knows" not only how fast it is currently driving but also how fast it should actually be driving. In the same way, it not only knows where it is currently parked, but it also knows how long it has been parked there and whether it is allowed to park there. By evaluating the corresponding data – whether in the vehicle itself or in the backend – violations of applicable regulations such as speeding or parking violations can be detected very easily and reported to the regulatory authorities or prevented from the outset through suitable technical measures in the vehicle. Road usage and parking fees could also be recorded and collected online in this way. In all these cases, there is the potential for cities not only to promote compliance with the rules and thus also road safety but also to avoid the expense of traffic and parking monitoring. The greatest technical challenge here would be the retrofitting of connectivity required for older vehicles.

At first glance, such massive encroachments on the personal freedom of action of vehicle drivers seem politically unfeasible, at least in democratically governed countries. However, a right to "free travel" is not really guaranteed anywhere. Nowhere in the world are there explicit rights for drivers to decide for themselves whether or not to keep to the speed limit or to park legally or even which route to take to their destination. The desire for freedom of action on the part of the individual is offset, on the other hand, by the desire for safety and a reasonable flow of traffic on the roads. Also, in the case of autonomously controlled vehicles (Sect. 4.2.2) at the latest, this question answers itself, because there is clearly no freedom of action for automated vehicles. Autonomous vehicles consistently comply with all regulations, regardless of the wishes of the passengers and can therefore also be ideally integrated into traffic management systems.

Based on the data collected in the vehicles, innovative services can be offered not only to their owners, but to all stakeholders in mobility. Such services can then, for example, also turn pure passengers without their own vehicle into loyal customers of a preferred vehicle brand.

4.4.3 Digitalization in Mobility Services

Even more than for the manufacturers and users of vehicles, digitalization holds great potential for the providers and users of mobility services. From public transport to classic taxis to car sharing and ride hailing, across all services offered, digitalization not only enables an expansion of the existing range of functions but above all an improvement in the *ease of use*, which leads to a significant increase in the acceptance and dissemination of mobility services and thus ultimately to a variety of new business models and offerings.

4.4.3.1 Public Passenger Transport

Those who want to use public buses and trains, but also airplanes or ships, can already have the best connection from their own location to their desired destination suggested to them in a few seconds via corresponding apps on their smartphones and, as a rule, also purchase the appropriate ticket online. This eliminates the tedious task of compiling the best connection from various timetables, as well as the need to go to a ticket machine or ticket counter.

All in all, however, public transport operators are still making far too little use of the opportunities offered by digitalization. According to a 2018 study by the German Transport Forum (DFV), less than half of the public transport companies surveyed are implementing digitalization projects. There is potential here, for example, in an online capacity utilization analysis, which could be used to optimize the planning of transport capacities and to show customers which buses and trains are particularly crowded at any given time. In a similar way to how route planning in a car navigation system takes into account current traffic flow data and avoids detected traffic jams, users of public transport could then be offered connections that avoid overcrowded lines.

At the same time, however, digitalization also harbors clear risks for public transport operators. Analogous to online vehicle sales, there is also the threat of losing the customer interface to third-party, private providers. Today, for example, mobility customers in Europe can already book and pay for bus and train tickets across cities and countries via the apps and websites of the company Trainline. This is of course highly practical, especially for cross-border mobile users: Instead of a multitude of different bus and train apps with separate logins and payment information, only one is needed. However, operators who are not listed there will have to distribute their services via other, but then shrinking, channels. This is high time, then, for public transport providers to take action. In Germany, for example, a number of transport operators have now joined forces in the *Mobility Inside* initiative in order to integrate singular services via a common platform and thus bundle public transport services for customers.

4.4.3.2 Car Sharing

It was only through digitalization that *free-floating* car *sharing, which* is now offered in an increasing number of cities, was able to develop alongside traditional *car* rental with stations at train stations, airports, and hotels. By connecting the central operator, users, and vehicles, the nearest vehicle can be identified, reserved, opened, and started via an app, and, at the end of the journey, it can be parked at any location within the contract area and locked and the journey paid for. In this way, carsharing allows mobility customers to make spontaneous and short trips quickly and conveniently.

Digitalization also brings a number of other advantages for the car-sharing operator:

* Reduction of the risk of damage:
  Accidents and damage to vehicles harm car sharing operators in many ways: They increase the insurance sum, take away the availability of the vehicle for other customers for the duration of the repair, and ultimately also affect the acceptance of the offer – after all, no one wants to book a visibly damaged vehicle. In order to reduce the risk of accidents, however, the operator can, for example, reduce the engine power for younger drivers or new customers via online access to the engine control unit and then – once reliability has been proven by accident-

free driving – gradually increase it again. At the same time, an online driving style analysis can be used to identify high-risk drivers and exclude them from the offer.

* Transparency with regard to damage and completeness:
In contrast to a rental car, which is usually checked for damage and completeness by a station employee upon return, an on-site inspection by the operator is not possible when parking a car-sharing vehicle. However, many customers forego the contractually prescribed visual inspection before starting their journey for reasons of time or convenience. However, sensors installed in the vehicle, such as RFID tags and RFID readers, can be used to remotely detect at any time whether movable vehicle components such as floor mats, warning triangle, spare wheel, or child seats are still on board. Damage to RFID tags can also be monitored; they then serve as a connected seal, so to speak. In this way, it can be monitored whether the first-aid kit has been opened or – via RFID tags attached to the underside of the vehicle – additional damage that is not visible to the driver taking over the vehicle can be detected, such as damage caused by carelessly driving over curbs. By evaluating the movement data during the parking period, it is also possible to determine whether the vehicle has been jostled during this time, and the previous user cannot therefore be held liable for the resulting damage. If appropriate cameras are available in the vehicle, the license plate number of the jostling vehicle can also be determined in this case.

* Optimal vehicle distribution:
One reason for the attractiveness of free-floating car-sharing services is the availability of vehicles close to the user's current location. A prerequisite for ensuring that as many people as possible who want to book a vehicle also find one available nearby is an optimal distribution of vehicles across the area of use. This distribution is changed by the use of the vehicles. For example, early on Sunday mornings, many of the vehicles are concentrated in the city center because users may have travelled there with them the previous evening but made the return journey in a taxi. The online recording of the vehicle locations then allows the operator to distribute the vehicles again in a targeted manner as required by the demand situation expected at the time.

- Transparency with regard to user behavior:
  Exactly what this optimal distribution looks like, i.e., when which vehicles should be available where, can be derived from a person-independent evaluation of vehicle usage data. AI methods can also be used to determine from this data which vehicle types are driven in a particularly dynamic manner, thus promoting wear and tear, and when is therefore the right time for maintenance work or resale.

In addition to cars, scooters, bicycles, and, more recently, e-scooters are also rented out in cities. Here, too, digitalization in the two-wheeler sector has led from rather cumbersome station-based offers to free-floating sharing, which is much more attractive from the user's point of view and can be used spontaneously and one-way depending on traffic situation, weather, budget, and mood.

4.4.3.3   Ride Hailing

The best example of the potential that digitalization holds for mobility, of how completely new market offerings can be created with relatively simple means and how providers that have been established for decades can suddenly be faced with problems that threaten their very existence, is the company Uber.

When it came to getting around in the city or the surrounding countryside without your own vehicle, taxis – if you didn't want to walk – have always been the only real alternative to public transport. This unique position, together with official regulation, has determined the quality and price level of taxis for years and certainly not always in the interests of the customer. In this environment, Uber, founded in 2009, has become one of the largest ride service providers in the world in just a few years, without even owning a single vehicle or hiring a single driver. An app for car owners who want to transport customers in their vehicle at agreed rates, and a second app for people with mobility needs who would like to use the services of these vehicle owners, was all that was needed. The app for the driver shows the location and status of the vehicle and shares relevant information about the vehicle and driver; the app for the customer is used to record the location and trip request, provide alternative drivers/ vehicles to choose from, and finally select one of them. Once they have arrived at their destination, the driver and the customer each confirm the end of the ride in their app, thus triggering the cashless payment – from which Uber, as the platform operator, retains its share.

In addition to the economic potential of such digital platforms, the example of Uber also shows the challenges that such innovative services pose for legislators. On the one hand, the legal framework for private ride-sharing services must be adapted to the centrally organized commercial driving service that has emerged from it, in order to ensure insurance protection for passengers, for example. On the other hand, a secure legal framework must also be created for drivers who work on a private basis without formal employment.

How Uber became one of the world's largest ride service providers purely through the development of two apps at its core, without owning a single vehicle or employing a single driver, is surely one of the greatest success stories of digitalization and illustrates its potential for success and disruption.

4.4.3.4   Breaking the Silos: Cross-Service Aspects

In addition to the influence on individual mobility offers shown above, there is considerable potential for digitalization in the cross-linking of different modes of transport. This is because the optimal connection from the user's point of view would often be a combination of different modes of transport and types of use, such as taking an e-scooter from home to the station, continuing by train, and then ride-hailing the last few kilometers to the destination in a car. For such a *modal split,* three apps and three accounts would still be necessary today, and, above all, there is no comprehensive route planning that takes into account all available offers and derives the optimal sequence for the requested route – ideally with criteria that can be prioritized individually, such as price, trip duration, or comfort.

As with the platform described above, which integrates different public transport offerings, such an intermodal platform shifts the supply chain of service provision, with all the associated risks for the existing players. The platform operator takes over the customer interface. Since trip planning, ordering, and payment are now only done through him, he now also owns the important and valuable customer data. The actual mobility providers are separated from their customers. Mind you, the customer has only advantages from this change, and to realize them, he only has to register in a new portal and will delete a few other apps in return.

4.4.4 Quality of Data-Based Services

The fact that the introduction of data-based functions and services not only creates completely new customer experiences, but also brings with it completely new criteria for customer satisfaction, is something that many manufacturers only became aware of after they had already rolled out the first of these functions. In the automotive industry in particular, this involves nothing less than a fundamental change in quality management requirements.

The first reason for this change is that, as already described, the functions of these services are not simply available after the vehicle has been handed over to the customer but must be actively ensured around the clock. They have to be "operated", and not only the legislator but above all the vehicle customer sees the operator responsibility for this quite clearly with the manufacturer. If a customer is stuck in a traffic jam due to a system error in the traffic flow display and misses an important appointment as a result, it is unlikely to matter to him whether the reason is maintenance work on the backend server, problems with the provider of the mobile Internet, or inadequate data from the supplier: What is not working in his eyes at this moment is a vehicle function, and, for this, in his opinion, the vehicle manufacturer is basically responsible. A historical and simply habit-based exception to this perception is the very first data-based service available in motor vehicles: the car radio. Although initially analogue and not digital, its function has always depended on whether and what the transmitting stations in the vicinity are saying. In contrast to the new services, however, no customer would think of blaming the vehicle manufacturer for false news or traffic jam information.

The second reason for the necessary change in the quality management of manufacturers is that, when customers use digital services, two product worlds that were originally experienced separately and with completely different expectations collide: the classic vehicle world and the new digital world:

* In the automotive world, customers expect maximum reliability, especially from safety-relevant functions such as brakes or steering but also from other systems. Manufacturers go to great lengths to ensure that the systems and their functions are safe during development and also formally assume liability for them when they are released. At the same time, the customer takes it for granted that the vehicle's range of functions will remain constant over its entire service life – apart from any retrofits. If a function nevertheless fails, a visit to the workshop for repair is required. If this happens within the scope covered by the warranty or manufacturer's guarantee, the repair is also paid for by the manufacturer.
* In the digital world, the same customers have had to live with limitations in reliability from the very beginning and have become accustomed to this over the years. The fact that the word processing program on the PC hangs, the messenger app on the smartphone crashes, the Internet access at home is "down", or the website of a provider is temporarily unavailable leads to annoyance but is more or less grudgingly accepted. The small print of the corresponding user contracts usually states that the provider accepts no liability for such cases. At the same time, however, digital products are continuously being developed. New releases with bug fixes, but also new appearance and functional enhancements, are made available to the customer at short intervals. The customer knows and expects that he always has the latest version of the service.

The challenge for vehicle manufacturers today is to resolve these contradictory customer expectations and to create and implement a coherent, brand-adequate, and at the same time economical quality image for the overall offering of vehicles and associated digital services. The first step in this process is a differentiated specification of customer expectations regarding the quality of services and functions. Relevant criteria here are reliability and availability but also up-to-dateness. Being unable to stream music in the vehicle for half an hour is probably less bad for the customer than receiving traffic flow information that is half an hour old at the crucial time. A brief interruption of the navigation display is annoying, but a brief unavailability of the e-call can prevent life-saving in the event of an accident.

On the other hand, there is high customer satisfaction potential in updating and expanding vehicle functions over the course of the vehicle's service life. A new design in the display or new options in the entertainment system not only give the customer the good feeling that his vehicle is becoming a little more modern or valuable each time, but it also shows him above all that his vehicle and therefore he as the owner are being looked after by the manufacturer in the long term.

Another important quality criterion for digital services is a consistent customer experience across the various available platforms. From the customer's point of view, the appearance, interaction, and function of a destination entry, for example, should be independent of whether it takes place in the vehicle via controller and display, on the smartphone via an app, or on the PC via a web portal. However, since these channels are currently managed and operated by completely different and often independent bodies within the organizational structures of the manufacturers, you still often have to look several times to recognize that different apps and websites come from the same manufacturer or belong to the same brand and unfortunately also enter the corresponding data twice.

Security against misuse by third parties also represents an essential quality aspect of digital services in the vehicle. The risks here range from unauthorized access to destinations or personal data to sabotage of vehicle functions. The technical implementation of the services must take appropriate precautions here; the secure design of the backend server already mentioned as a firewall to the Internet is one of these.

One aspect of customer satisfaction with digital services that is still rarely satisfactorily resolved by vehicle manufacturers today is customer care. Complexity has increased here, of course: The malfunction or unavailability of a service, because of which the customer contacts the hotline, for example, can no longer be rooted solely in the vehicle but in all the subsystems involved: Activation of the services by the operator, compatibility of the different versions of app with smartphone, and program with PC (in each case including the versions of the operating system), availability of backend, cloud, and mobile Internet are all possible causes; a fault analysis is much more complex and requires much better systemic support. People and systems used in customer support for digital services must have significantly more knowledge and solution competence than in the classic vehicle hotline.

4.4.5 Legal Aspects of Data-Based Services

When talking about legal aspects in connection with data-based services, the protection of personal data is usually in the foreground. However, in addition to the legal requirements for data protection, there are also regulatory requirements as to when and which data must be collected and, if necessary, also forwarded.

4.4.5.1 Who May Process Data and Under Which Conditions?

From a purely technical point of view, vehicle manufacturers and mobility service providers could earn money with the data generated or that can be generated in their vehicles in a wide variety of ways and with a wide variety of customers. If, for example, every vehicle occupied by at least four people that has been driving on the highway for more than 3 hours without a break were to be sent an appetizing hint about the next restaurant of a certain chain, the operator of the restaurant chain would certainly be willing to pay a not inconsiderable price for it. However, one of the obstacles to the implementation of such use cases is the sometimes rigid data protection laws, which have been tightened worldwide over the past few years in light of the digital transformation.

Who is allowed to process personal data, i.e., collect, record, store, modify, transmit, or evaluate it? Since May 2015, this is regulated in Europe by the General Data Protection Regulation (GDPR), which is valid in all member states of the European Union, has emerged from a multitude of national and in some cases even regional legal regulations, and is the model for the corresponding legislation of many non-European countries and economic areas.

For the area of mobility that is relevant here, it must first be clarified which data is actually considered personal – first and foremost, of course, customer data, such as that recorded and stored during personal registration in online portals or when concluding a contract. Here, the customer's consent to the storage of his data is of course required, but it is also clearly bound to a specific purpose. However, even the technical data that appears at first glance to be purely vehicle-related and independent of the person, which is collected by a vehicle during its use via sensors and control devices, can be clearly assigned to the responsible vehicle owner via the chassis number and is therefore legally handled as personal data. This includes, for example, vehicle speed, usage statistics, condition data of wear parts, fault memory entries, or, in particular, the vehicle position.

The central requirement of the GDPR and thus a prerequisite for any processing of personal data is the unambiguous consent of the person to whom the data relates. Obtaining consent can only be waived in very specific circumstances, such as when the collection of data is a necessary condition for the performance of contractual obligations, is necessary for the preservation of the life of individuals, or is in the public interest. For the field of mobility, this means that an opt-in, i.e., a declaration of consent to data processing, is required in particular for commercial data-based services such as an electronic driver logbook or driving style analysis. However, when it comes to the performance of a contract, such as the recording and storage of the arrival and departure time in a car park; safety-critical services, such as the reporting of an accident including its location to the emergency call center; or services in the public interest, such as the storage of video recordings from surveillance cameras in public transport, personal data may also be processed without consent. However, the limits as to when these conditions are deemed to be met are drawn very narrowly by the legislator and the courts. For example, the collection of data to avoid a breakdown is not considered a measure to preserve human life and is therefore always subject to consent.

In addition to the aspect of data protection, the question of who actually owns the data collected in the vehicle or from the user is also relevant. In the European Union, the current legal opinion is that data are not things and therefore there can be no ownership of data – not even on the basis of data protection law or copyright. However, against the backdrop of the economic potential inherent in the use of data in the field of mobility, a possible ownership right to data is being discussed again and again.

The fact that the legal situation can be completely different as a result of small changes in the boundary conditions is shown by the following four data processing scenarios using the example of the measurement of the engine oil level, which at first glance appears to be completely harmless:

- Today's passenger cars have sensors for measuring the engine oil level and a system for monitoring it. If the oil level falls below a specified limit, the driver receives a corresponding message in the display. In this case, data is collected and monitored but not stored. The processing is necessary to implement the contractually guaranteed on-board oil level monitoring function. Consent to the data collection is therefore not required, and no vehicle customer would probably object to the collection and evaluation of the data.
- The case is somewhat different if the manufacturer is automatically informed when the engine oil level is low, so that the manufacturer can make the vehicle owner an appropriate offer for engine oil, including refilling, at the right time (the corresponding type of oil then happens to be on offer ...). This type of data processing is clearly aimed at the preparation of a commercial offer and therefore requires consent. Moreover, it may exclude other suppliers of motor oil, such as service stations or independent repairers, from competition.
- The situation is different again if the vehicle not only continuously collects the oil level data but also stores its history on-board. If, for example, the vehicle suffers engine damage at some point and is taken to the workshop for repair, the dealer can read out the stored oil level history there and use it to check whether the vehicle owner has fulfilled

his operator obligation and topped up the engine oil regularly. If, on the other hand, the vehicle has been driven over longer distances with too little oil, the owner may forfeit his warranty claim as a result. Consent to the use of data is also mandatory in this case, but the owner will carefully consider here whether he agrees to the data transfer or not.

* In the fourth scenario, the data is also collected continuously and transmitted to the manufacturer. The manufacturer collects the engine oil level histories of all its vehicles in the field and sells this data collection to engine oil manufacturers. The vehicle owner is not harmed by this, but since the manufacturer makes a profit from the data collected from his vehicle, he may well consider making his consent to the use of the data contingent on a share of that profit.

The last scenario also illustrates the important difference between raw data and refined data. Raw data refers to the data as recorded by the vehicle sensors, such as the individual oil level measurements of the vehicles under consideration. If this data is evaluated and processed into meaningful information, such as usage statistics for certain types of oil, it is referred to as refined data. If this data can then neither directly nor indirectly be assigned to individual persons, consent is no longer required for its further processing.

4.4.5.2   When Do Data Have to Be Collected?

In the majority of cases, the focus is on the question discussed above, under which conditions data may be collected and further processed. In addition, there are two different aspects under which the collection and disclosure of data is even required by law:

On the one hand, public authorities have an interest in vehicle usage data and may request this from the manufacturer. This may involve information on individual vehicles, as in the case of criminal investigations by investigating authorities but also statements relating to the entire vehicle fleet of a manufacturer or parts thereof. For example, in the interests of consumer protection, authorities would like to know how many vehicles in the fleet have a particular fault, so that they can assess whether these are individual cases or a systematic fault that then requires a recall. In addition, the bodies charged with drafting the relevant legislation are also interested in data on vehicle usage patterns. For example, BEVs and PHEVs registered in China must continuously record a specified set of data and transmit it to the relevant authorities. This data is then used to determine, for example, whether drivers of PHEVs really do minimize emissions by driving as much as possible on electric power, thus justifiably benefiting from tax breaks, or whether PHEVs are only bought because of the associated tax advantage but are primarily powered by the internal combustion engine.

The second aspect is the aforementioned protection of competition, which independent workshops and other service providers in particular demand. In the EU, for example, a *block exemption regulation* to ensure fair competition in aftersales already stipulates that vehicle manufacturers must disclose their repair and maintenance data. Using corresponding IT applications such as Repair and Maintenance Information (RMI) from TecAlliance, every workshop cannot only access repair data such as tightening torques but also order the relevant spare parts and special tools. However, if repair and service requirements are now determined by the vehicle and transmitted "over-the-air" directly to the manufacturer, the latter will have a clear advantage over the competition when it comes to submitting corresponding service offers. From 2020, remotely accessible service data must therefore also be made available to all service providers in the EU.

Today, users think twice about whether and for what they disclose their personal data. But anyone who offers real added value and shows the necessary care and transparency in handling the data entrusted to them will always get their consent.

4.4.6 Digital Culture: More than Just Full Beards and Sneakers

Anyone who has worked in interdisciplinary teams knows how much not only the mindsets and working methods but also values and personal style, right down to clothing, expressions, or hobbies, can differ between engineers, business economists, or lawyers, for example. Nevertheless, as part of a cross-company and cross-industry *corporate culture*, there is usually at least a common understanding of the roles and tasks of each individual and, above all, of the work processes that goes beyond formal documentation. Such a common understanding is then expressed, for example, in the statement "That's just the way we do things".

In contrast, what we perceive today as *digital culture* is a common mindset across all industries among people who – in the broadest sense – develop, operate, or even just consciously and enthusiastically use IT systems. Far more than existing corporate cultures, it is a general positive, progress-affirming, and IT-savvy attitude in work and private life in which above all many traditional and established things are critically questioned with regard to their suitability in the digital age and, if necessary, thrown overboard.

New technical solutions and changing customer expectations are leading to more or less significant changes for all partners involved in existing value chains: Many established business areas (not only the manufacture of combustion engines and the associated components) are being called into question in their existing form, while new opportunities are opening up in other areas, such as the provision of mobility services. It is becoming clear that these opportunities are not being exploited by the long-established players in the industry but by new digital players that are entering the competitive arena of mobility providers without historical ballast, with lean processes and a fresh corporate culture. Also, the competition in this arena is not just about customers, customer interfaces, and market share but – much more existentially – also about qualified and motivated employees. Those who decide too late to make the change may not be able to implement it, if only because they cannot get the "digital" employees they need on board. However, it is precisely these employees who look to potential employers for different benefits than was previously the case. A company car, an office of one's own, and bonus payments dependent on company profits are no longer sufficient here; soft factors such as a positive corporate culture, a cooperative management style, work-life balance, or simply a sense of purpose in one's own work are at least as important for digital employees.

This digital transformation has been taking place for years and not only relates to the use of new (digital) technologies and agile working methods in particular but is also taking place in the middle of society, in the personal behaviors, and expectations of all stakeholders. Even if the "older" generation, which still has the majority in the upper management of many companies and makes the decisions there, does not really want to admit it, a considerable part of professional, social, and private life now takes place on the Internet and via smartphones. Not only behavior patterns are shifting but also personal priorities and values, thus shaking up the old world views. In terms of potential income, bloggers or e-gamers, for example, are sometimes to be taken more seriously as a profession than the "proper" university studies or "solid" vocational training so readily demanded by parents. The right smartphone is much more of a status symbol among the younger generation than the right car. Anyone who wants to successfully offer mobility in the future, whether as a product or a service, must adapt to this fundamental change, not only among their customers but also among their employees and managers. Even if older politicians in particular always have a hard time with the digital culture in public, the digital transformation has also long since made its way into politics and administration. In many places, people who are not connected can neither submit applications to the authorities nor pay for a parking space.

4.4.6.1 "Old School": The Way It Is (Or Used to Be) Done

As in many other areas of goods production, the primary business model in the automotive industry to this day is to take customer requirements, implement these requirements optimally in a product – namely, the motor vehicle – and then sell this to the customer in as perfect a quality as possible. Even though the associated processes have been continuously developed and improved, the underlying business model has remained the same for over 100 years:

- The *development of* a vehicle follows a highly complex plan that has been optimized down to the last detail over the years and controls the interaction of all the specialist departments involved. Within the specified period of time, the design is developed and validated with an increasing degree of concreteness until the resulting vehicle design meets all agreed and expected requirements and the vehicle can be produced in series – i.e., the desired concept quality and readiness for series production are achieved.

- From this point on, *series production of* the vehicle begins in one or more plants, i.e., as smoothly as possible, like clockwork, up to several hundred times a day. The aim of the underlying production process is to produce the vehicles exactly according to the specifications of the design, i.e., in optimum production quality. The characteristics, functions, and quality of the vehicle are thus usually defined at the moment the vehicle leaves the factory and is transported to the dealer or at the latest when the dealer hands it over to the customer.

The manufacturers' top priority in this process chain is to bring a vehicle onto the market that is as perfect as possible – which includes not only development and production but also the sales process. Even today, the development of a vehicle from the product idea to the achievement of series production readiness can take up to 7 years and cost hundreds of millions. From this point on, however, it is also complete, and the subsequent *series production support and further development* only involves comparatively minor design changes – for example, to upgrade and update the product substance, as in the facelift common from the third year of sales onwards, to reduce manufacturing costs and, to a not inconsiderable extent, to rectify quality problems.

In comparison, the product usage phase is given much less attention. After the vehicle has been handed over to the dealer, it is no longer the sales staff but the aftersales staff who are responsible for the customer. These take care of scheduled maintenance work such as oil changes, accessory offers such as winter wheels, and – if necessary – the repair of accident damage and technical faults. Apart from that, the manufacturer normally only gets in touch with the customer again when the purchase of the successor vehicle is pending; continuous customer care during the use phase is the exception and, if at all, then reserved for the top price segment. Yet, at the latest with the advent of digital services, it is precisely this phase that represents the greatest lever for active customer loyalty and competitive differentiation beyond the product substance. Fewer and fewer customers remain loyal to a brand out of pure enthusiasm for the product.

4.4.6.2   Agile: The Magic Formula from Software Development

In contrast to this approach derived from hardware development, a completely different path was taken in the development of software right from the start. Since software products can be duplicated, delivered, and installed with much less effort than hardware, it is possible to initially deliver them at a much lower level of maturity, but also much faster, and to fix any errors or problems very quickly with new releases. The agile guiding principle "Better done than perfect", attributed to Facebook founder Marc Zuckerberg, succinctly sums up this way of thinking and proceeding. Even though Bill Gates and Microsoft probably didn't even know the term "agile" in the 1980s, anyone who worked with the first versions of Microsoft Word back then probably remembers today first and foremost the many functional errors and program crashes of the first releases but also the steady increase in functionality and reliability in the period that followed. Obviously it did not harm the spread and success of Word. The probably fictitious, but often quoted debate between Microsoft founder Bill Gates and the then General Motors boss Jack Welch, in which Gates complains how cheap and economical cars could be if manufacturers like GM would develop as fast as Microsoft, and Welch replies how unreasonable and life-threatening cars would be if they had the quality of Microsoft's software, illustrates the peculiarities of the two different approaches.

Agile working methods have made it possible in software development to quickly bring the first usable results to the market and thus to be able to react flexibly to the market and competition at any time. In order to be able to transfer this effect to other areas of product and service development, the corresponding methods were adopted and adapted to their requirements under the collective term *agile development*. One of the best known and most widespread agile methods is Scrum. Here, cross-functional teams solve the development task step-by-step and, above all, in a self-organized manner over several sprints of equal length. Agile development differs from the established, sometimes strictly hierarchical and lengthy development processes of the automotive industry, for example, not only in its self-organization but also, among other things, in the rapid delivery of a *minimum viable product (MVP)*, i.e., a minimal solution that is by no means perfect but just about acceptable to the customer, which is then improved step by step with new releases. Even here it becomes clear how radical such approaches were initially from the point of view of established work structures and processes, especially in the automotive industry. Whereas the "minimum" in MVP for safety-critical functions such as steering or braking means 100 percent reliability, as before, compromises are more possible at the beginning for pure comfort functions.

If digital functions are to be integrated into established, complex products such as a car, for example, the challenge is to combine the existing, typically mechanics-oriented and thus comparatively slow development processes with the fast, agile processes of software development. Technically, this requires the greatest possible decoupling of product hardware and software as well as standardized, backward-compatible interfaces. This is the only way to ensure that, for example, an electric vehicle purchased today will still be able to charge at the public charging stations that have been further developed by then or connect to the latest version of smartphones at that time.

4.4.6.3   Clash of Cultures: Two Worlds Meet

In addition to the technical aspects of this joint development of hardware and software, another and presumably far greater challenge is the clash of the different values and cultures that the representatives of the two "camps" bring with them. The agile mantra "better done than perfect" is obviously diametrically opposed to the guidelines of the classic understanding of quality in the automotive industry, such as *"quality first"* or *"zero tolerance"*. Anyone who develops and offers data-based services in a

2-week cycle today often only has to shake their head in disbelief at the years-long development times for the complete vehicle with the optimization cycles and rigid release processes that take months due to the construction and testing of hardware prototypes, while, conversely, from the point of view of many seasoned automotive developers, the agile working methods contradict all the learned and sacred principles of the art of engineering and exhibit almost irresponsible anarchic traits.

However, these cultural differences go far beyond practical working methods and apply deep into the core, into the companies' understanding of leadership. On the one hand, there is the classically unassailable management decision based on the personal and, of course, not always validatable expertise of the responsible managers. Such decisions may well be based on technical knowledge, but they can also be based on purely subjective perceptions, such as "Everyone at my golf club wants electrically adjustable interior mirrors in their cars, we absolutely have to introduce them!" If no one dares to contradict this despite knowing better (and this is still common in many companies), the personal opinion becomes a company decision, possibly with the corresponding consequences.

In the digital society, however, facts count more than personal opinions. Just as the smartwatch continuously tracks one's heartbeat and the networked scale continuously tracks one's BMI and thus assesses one's personal well-being, the results of data analytics are generally trusted more than even the management board. In order to find out, for example, how customers actually behave and which offers they really find good or bad, sales managers are not asked here, but corresponding data is collected and evaluated in the field. Among the "digitals", for example, the slogan "Don't let HIPPO decide, let data decide!" applies, whereby "HIPPO" here stands for Highest Paid Person's Opinion. Yet the realization of how important data is to business decisions is truly not new: W. Edwards Deming, one of the leading thinkers on quality management of the last century, is credited with the statement "Without data, you're just another person with an opinion", which is consistently perceived as radical and confrontational by board members and business leaders.

In many companies, this kind of thinking is seen as radical and dangerous, since its implementation ultimately means nothing less than taking away the decision-making power of managers and their committees. Giving up processes and structures that have been practiced for decades and, in retrospect, have been highly successful for those who have made it to the top of the company is just as difficult for one side as it is for the other to accept that some product goals, such as safety requirements, are non-negotiable and must be achieved 100 percent at the first attempt. Integrating the old world and the new world not only in terms of content, but above all socially and culturally, is the critical task of corporate management in the digital transformation.

4.4.6.4 Digital Transformation in Society

On the one hand, digitalization involves the technical change brought about by the mass and intelligent interconnection of "things" on the Internet, especially their sensors and actuators. The change relates to products and services, including the processes for their production or provision. On the other hand, digitalization is also reflected in the new forms of offerings already described, the *digital business models*. In the area of mobility, *e-commerce*, for example, in the form of online car sales, subscription *models* such as a traffic flow information service that can be subscribed to, and *pay-per-use offers* such as car sharing are particularly relevant here. All in all, the focus of all digital business models is no longer on selling a previously developed and produced product but on operating products as a service: no longer CDs and DVDs, but streaming; no longer digital road maps, but intelligent route planning in real time; no longer cars, but mobility.

These new business models are so promising not least because digitalization has triggered a profound change in all areas of society beyond the business world in recent years. The widespread use of smartphones across all social classes, all age groups, and all geographical areas has made a significant contribution to this. Hardly any device has ever so radically changed the lives of so many people in so many different areas in such a short time as the smartphone and the personal and mobile connectivity it makes possible. The fact that everyone, no matter where they are in the world, can be in touch with each other and exchange data at any time has led to technical, business, social, and political changes of a magnitude not yet foreseeable today.

Anyone who was born before 1990 and thus consciously experienced this change as an adult will usually take a close look at the changes compared to what they were used to and will definitely evaluate them critically. What is the quality of the news and information available on the Internet? Is it really only advantageous to always be online and thus available? If you spend more time on your smartphone, what other things do you no longer have time for? The *digital natives* who grew up with smartphones and tablets, on the other hand, lack this experience as an alternative; at most, they take note of it in old films and from stories told by their parents or even grandparents with smirks or disbelief: Did you really have those big telephones with cables attached and pay 3 euros per minute for a call from Germany to England? Did you really have those thick department store catalogues and wait 6 weeks for delivery? Did you really have to buy tickets at ticket machines and pay with coins?

Regardless of whether one has grown up with digitalization or learned it in later years of life, in the digital society, consumers have significantly different expectations of products and services than was the case just a few years ago. On the one hand, anyone who wants to purchase a product or use a service here expects to be able to do so anytime and anywhere. Regardless of whether they are sitting in a café or at home on the sofa, they want to be able to obtain information and advice around the clock and book or order at any time. If they have to go to a shop to do this, they do not want to have to observe closing times there either. Restricted service hours are generally considered unacceptable. A mobility service that is not available between ten in the evening and six in the morning has no chance among *Digitals*. Secondly, availability is expected to be as immediate as possible. Even though there have been times when car buyers have accepted waiting up to 5 years for their car to be delivered without batting an eyelid, in times of Amazon Prime and same-day delivery, even 5 weeks of delivery time is perceived as interminably long, and those who want to book a ride-hailing service see even 10 minutes of waiting time as absolutely borderline. Also, anyone standing in front of a sharing e-scooter in a foreign city wants to drive off with it immediately. If registration takes longer than 2 minutes, the customer is on to the next provider.

So anyone who wants to do business with born or become *Digitals in the* long term should take this change in expectations into account and also adapt to changed boundary conditions beyond that:

- Declining loyalty
  Those who are used to buy even the daily necessities online will not limit themselves to the offer of the local dealer when buying a car. The car-sharing provider or the vehicle brand will be changed just as quickly as the new partner is found on Tinder or Parship or a new internet provider or wine supplier is identified on a comparison portal. "One Click In" also means "One Click Out".
- Wide range of information
  Via the Internet, customers cannot only obtain comprehensive information about a particular offer, including the associated customer experiences and alternative offers from the competition; reports and information about the provider itself are also obtained and can influence the purchase decision. Digitals are value-oriented and are not only alert but also sensitive. Anyone who offers a great product but is associated in Internet forums with the exploitation of local workers or the causing of environmental damage is quickly out of the running.
- Feedback culture
  Receiving feedback about oneself, one's own product, or one's own service is always valuable. While it used to take a certain amount of effort to obtain feedback, for example, through customer surveys, in the digital culture it is given unasked, quickly, unfiltered, and, above all, accessible to everyone. People and companies alike have to be able to cope with the fact that the experience of a dissatisfied customer can be read worldwide just minutes later in the relevant Internet forums and networks. Anyone affected by this also quickly learns that in most cases it is far more effective to swallow this feedback without complaint, even if it is perceived as unfair, and to derive improvement measures from it than to argue in front of the whole world whether and why the criticism was justified or not.

- Enlightened relationship with personal data
  When a small amount of personal data was to be collected as part of a census in Germany in the 1980s, a wave of protest arose. Opponents of the census did not have a specific reason for their opposition but were generally opposed to having to reveal information about themselves. Many of those who protested the census at the time or refused to provide the requested information have, 25 years later, voluntarily and unreservedly given up personal information on a much larger scale just to collect points on Payback or receive personalized offers from online vendors. In the meantime, after a flood of spam mails and data abuse scandals, people are once again paying much more attention to the protection of their personal data and are carefully weighing up for what purpose they are prepared to pass it on and for what purpose they are not.
- Use instead of own
  Music and film lovers who subscribe to streaming services for the first time have an almost classic digitalization experience: The collection of CDs and DVDs that they have nurtured and cared for over years and decades sinks into complete insignificance virtually overnight. Using digital services, in this case streaming songs and videos, proves to be not only much cheaper than owning them but also much easier and more convenient. I don't have to limit myself to the music and movies that are on my shelf but can listen to and watch whatever I feel like. This applies analogously to mobility. If you don't need a car every day and don't need the same car every day, on-demand services are often cheaper and more flexible. This change is clearly visible in the changing value of possessions. The CD and DVD collection have clearly lost their former importance as a status symbol today; the own car is just on its way out.

So while society is already deep in the digital transformation, many established companies are clearly still struggling with it. The reason for this is primarily managers and employees who cling to the processes and values that were successful in their eyes in the past and that ultimately put them in the position they are in today. From this perspective, digital change means having to give up one's own recipe for success and betraying one's own values. It is thus perceived as a threat and must be stopped or even fought. It is obvious that such a mindset also lacks the courage to make decisions for change and the long-term preservation of competitiveness.

Digital Culture combines an affinity for new technologies with a work and lifestyle based on transparency, mutual respect and sensible prioritization. Digitals are therefore also often at odds with traditional corporate cultures, for example when they trust data more than the personal opinions of managers.

5

Mobility as a Service
What Alternatives to Owning a Vehicle Will There Be in the Future?

5.1 The Mobility Classic: The Own Car

5.1.1 The Own Car as an End in Itself

For many people, owning their own car was and is not just a professional or private necessity but a value and a goal in itself. The ability to drive whenever you want and wherever you want has made the car an epitome of personal freedom from the very beginning. Through the costs associated with its acquisition, as well as the emotions expressed and absorbed by different types of vehicles, it has also become a significant means of personal self-expression and a symbol of social status. In many areas and societies, it still represents the highest level of individual mobility and is seen as an indication of the place its owner is accorded within society: Those who have enough money or are old enough can drive their own car and are no longer dependent on buses, bicycles, or mopeds. In China, where in recent years many people have switched from bicycles to electric scooters and then to cars in a very short time (a development that previously took about 10 times as long in the West), this effect could and can be observed almost in fast motion. The importance of the car, which goes far beyond its function as a means of transport, is also illustrated

J. Weber, *Moving Times*, https://doi.org/10.1007/978-3-658-37733-5_5

worldwide by the fact that in many private households the purchase of a vehicle is only surpassed in terms of investment volume by the possible purchase of a house.

The social status associated with owning a car is a complex issue and, among many other aspects, depends above all on the respective cultural area and social environment. The primary factors here are type, class, and brand and the associated price position of the vehicle. The relationship between price and status is by no means linear, and the effect intended by the owner and the perception granted by society are also sometimes far apart. For example, drivers of SUVs in city centers today often experience rejection or even aggression instead of the recognition or even admiration hoped for by the size and presence of the vehicle, while the odd used small car can certainly become a sympathy catcher for its owner.

At the same time, the fact that the more vehicles have to share the available roads and parking spaces, the fewer advantages there are to using one's own car has already led to a shift in mobility behavior and the associated perception of status in international metropolises at an early stage. The subways and buses of New York, London, or Tokyo have long since been used by more than just people waiting to finally be able to afford their own car. Employers are now increasingly offering flat-rate mobility allowances instead of the company cars that used to be the norm, and employees can then decide for themselves exactly how they want to use them. Social acceptance of public transport as an alternative to owning a car can vary considerably even between cities in the same cultural area; it is much more pronounced in New York, for example, than in Los Angeles. There are also clear differences between urban and rural areas: while public transport is increasingly becoming an alternative to the private car in many cities, hardly anyone voluntarily prefers to take the bus rather than their own car for the journey from one village to a neighboring village.

5.1.2 Alternatives to Owning a Car

For a long time, however, anyone who wanted to get from A to B without their own vehicle essentially had three alternatives: to use local and long-distance public transport, to be driven by taxi, or to rent a car. Additional services, such as car-sharing agencies for trips from city to city, privately organized sharing of private cars, or the use of buses chartered by companies to take employees from factories to work or customers to large shopping centers, represented rare and narrow market niches in the mobility market.

However, this limited offer has changed significantly within the last 10 years. Between the private car or two-wheeler, the relatively inexpensive but inflexible public transport system, and the comparably expensive rental car, a large number of new offers have become established; mobility customers can now choose the optimal solution for their personal situation in a much more differentiated manner. Three parallel effects are responsible for this significant market change:

- New options through digitalization: The ability to submit one's mobility wishes and call up available offers at any time and from anywhere via smartphones enables service providers to launch attractively priced mobility offers tailored to customers' personal needs.
- Increasing dissatisfaction with the current situation: Owning and using one's own car is increasingly inconvenient, especially in cities, due to the deteriorating traffic and parking situation, rising costs such as taxes or tolls, and regulatory constraints such as access restrictions. At the same time, overcrowded trains and buses as well as unreliability are also making public transport increasingly unattractive in many places.
- Changing values: Especially among younger people, the importance of the car as a status symbol and the basis of personal freedom is increasingly declining.

So when are people prepared to give up their own car? The decisive factor is the extent to which the attractiveness and social esteem of alternative mobility options increase and, at the same time, the "pain" associated with using one's own car increases. For example, anyone who lives in a small town and has never seriously (let alone regularly) been stuck in a traffic jam or had to search for a free parking space will hardly ever have really thought about alternatives to owning a car. In many cases, professional or private necessities oppose or categorically rule out the alternatives. For example, someone, who frequently has to be mobile at short notice, who as a sportsman or craftsman needs special equipment that is to remain in the car, or who regularly travels from places or to destinations where no alternatives are available, will put up with many inconveniences before giving up their own car. Certainly, for many people, owning their own vehicle – even if not absolutely necessary – will continue to represent such a high value in itself that they will be willing to pay a price for it, no matter how high.

What makes the decision to do without one's own car easier are the new, digital mobility services, which sensibly complement the existing, timetable-based public mobility. Apart from switching to alternative vehicles such as bicycles or scooters, there are two main alternatives:

• Car sharing, that is, driving yourself in someone else's car
• Ride hailing, that is, getting a ride in someone else's car

Since this usually involves giving up familiar and, above all, cherished behaviors, this process of weaning oneself off the car usually takes place in two stages: The first step is the use of car-sharing services. Although one no longer has one's own car, one still drives oneself "as in the past". The use of ride-hailing services then represents the second stage and, with the renunciation of driving oneself, requires a further letting go.

No matter where in the world: in the countryside and in small towns, the private car will still be the predominant means of transport in ten or twenty years' time. So in that respect, car manufacturers have little to worry about.

5.2   Car Sharing: Driving Yourself in a Borrowed Car

The statistics show a clear and sobering picture: In Europe, privately owned vehicles are driven for an average of about 1 hour every day, and they remain parked for the remaining 23 hours. What's more, parking usually involves costs as well. In many city centers, the rental price for a parking space is now in the order of magnitude of the leasing rate for a mid-range car. Soberly considered, the vehicle owner pays just as much for the nonuse of his vehicle as he already pays for its use – an economic insanity, which is aggravated by increasing usage costs.

Borrowing a vehicle on demand, on the other hand, offers the possibility of really only paying for the time of actual use. Depending on the individual situation, car sharing is therefore in many cases the more economical alternative, not only to a second car but in many cases also to a first car.

5.2.1 Offers and Business Models

5.2.1.1 Business-to-Consumer (B2C) Car Sharing

In *B2C car sharing*, the operating company owns a fleet of vehicles and makes them available to its customers against payment. In order to hedge the operator's risks, to comply with legal requirements and finally to simplify the usage processes, customers have to register as members before using the service. This registration includes the assurance of identity, a valid driver's license, and a payment option.

Rental Car

The business model of classic car rental companies such as Enterprise, Hertz, or Sixt is that customers register their vehicle requirements in advance, if possible in the form of a reservation (otherwise the range of vehicles available can be very limited); an employee of the car rental company hands over the vehicle in question to the customer at one of the desired stations, washed and with a full tank of petrol, and the customer usually returns it to the same station after completing the journey, where it is checked for completeness and damage by other employees. The customer is charged by the day, with significant price reductions for trips lasting several days or weeks. Trips where the vehicle is returned to a different station than it was picked up from are possible but significantly more expensive.

While rental cars are still the usual mobility solution for business and holiday trips due to the lack of alternatives, they are highly unattractive for types of use such as regular commuting or spontaneous and short city trips due to the commitment to stations, the time-consuming reservation, and return processes as well as the daily billing.

Station-Based Car Sharing

The next level after rental cars, which is already a good deal more flexible and attractive, is *station-based car sharing*, which has been available for some time. The vehicles still have to be picked up and returned at fixed stations, but the process is significantly simplified and usually runs without interaction with an employee via access card or smartphone – which saves costs and enables significantly lower prices. Charges can usually be made per hour. Fuel costs are included in the price. Should it be necessary to refuel during the course of the journey, this is done by the customer himself at contract filling stations, where he then pays with a fuel card enclosed with the vehicle.

The fact that the vehicles are always returned to their station makes station-based car sharing comparatively low-effort and low-risk for the operator. Due to the synergies with the core business, the most important providers come from the field of classic car rental, examples being Enterprise Car-Share or the pioneer and market leader Zip-Car, which was acquired by Avis in 2013. However, Hertz on Demand shows that the business model is not always successful: Hertz's car-sharing service was already discontinued in 2014 due to inefficiency.

In addition to car rental companies, some vehicle manufacturers have also taken up car sharing as a new business model, such as General Motors' Maven service or BMW on Demand, which is only available in Munich. Understandably, only vehicles of their own brands are offered, which makes car sharing not only a new business segment but also a marketing tool: The offer is based on the intention and hope that the car-sharing customer is so enthusiastic about the booked vehicle that he will sooner or later buy an identical or similar model. However, such a "paid test drive" is clearly in contradiction to the role of car sharing as an enabler of the renunciation of one's own car, which is especially intended by municipalities and is therefore viewed quite critically by them, particularly with regard to a possible subsidy or support.

Another group of providers for station-based car sharing are the operators of long-distance public transport, who can thus position their services as an alternative to driving one's own car, even for decentralized destinations. In Germany, for example, a journey by train is an option for significantly more customers if even remote destinations can be reached comfortably from the arrival station using vehicles from the in-house car-sharing service Flinkster. In this case, the fact that the rented car has to be returned to the station at the end of the trip does not mean any additional effort for the customer.

Free-Floating Car Sharing

For spontaneous city trips, however, the hourly billing and the station commitment of station-based car sharing represent a real obstacle. Its use is ruled out from the outset if the vehicle in question has to be returned to the pick-up station at the end of the trip.

In this case, the possibility of recording the exact locations of users and vehicles via smartphones and connected vehicles makes a much more practical alternative possible: *free-floating car sharing*, which has been offered in many cities for some years now, in which the available vehicles are distributed across the city and the requesting user is then shown the ones that are closest to him. The desired vehicle is located via the provider's app, opened and made ready for use; completeness, cleanliness, and integrity are queried via a menu in the vehicle and confirmed by the user. After reaching the destination, the parked vehicle is locked via the app, and the trip is billed to the minute using previously stored means of payment.

The start and end of the journey can therefore be anywhere within a predefined area of a city or conurbation. The acceptance of free-floating car-sharing services stands and falls with a distribution of the vehicles over the operating area that corresponds to the needs of the users. In the event of major deviations between the target and actual distribution, the vehicles must be redistributed here either individually with drivers or by loading them onto vans, for example, if too many vehicles are concentrated in the city center early on a Sunday morning and there are no vehicles in the residential areas on the outskirts of the city.

The cost of fuel or electricity is also included in the price of free-floating car sharing. When the available vehicles are displayed in the booking app, their current fuel level or battery charge status is also shown so that the user can assess whether this is sufficient for the planned trip. The vehicles are then refueled, charged, or washed by the customers themselves, who in return receive free minutes from the operator for using the service. This is a clever trick that saves the operators the corresponding personnel costs: In a very short time, user groups have been found who specifically only look for vehicles that need to be refueled or charged and thus finance their individual mobility.

A significant cost factor for the operators of free-floating car-sharing fleets are the high expenses for maintenance and repair. On the one hand, the vehicles are used much longer and more intensively within the same period of time than privately owned vehicles and are therefore naturally subject to greater wear and tear. On the other hand, the lack of a condition check by the provider when the vehicle is handed over, the fact that the fare is independent of consumption and the feeling of being unobserved during the journey quite obviously tempt many users to a less careful driving style and to careless and sometimes even deliberately damaging handling of the vehicles. Both wear and tear and the frequency of accident damage are significantly higher for sharing vehicles than for comparable privately owned vehicles.

Today, free-floating car-sharing fleets are usually operated by vehicle manufacturers – such as FREE NOW as a merger of Daimler Car2Go and BMW DriveNOW or We Share by Volkswagen. One of the main reasons for this is the marketing expectation of the manufacturers, described above, that the customer will get excited about buying a vehicle while driving a shared car. For this reason, some of these operators are also toying with the idea of allowing the user to purchase the vehicle as a used car at the end of the journey directly from the menu with a single click so that he can continue driving "his" vehicle straight away. Another reason that makes the operation of car sharing services particularly attractive for vehicle manufacturers is that they can carry out mainte-nance and repair work on the vehicle fleet through their in-house or con-tract workshops at comparatively low cost.

5.2.1.2   Peer-to-Peer (P2P) Car Sharing

The business model of *P2P car sharing* is based on the fact that private individuals are willing to let other users use their own cars temporarily and for a fee. Via websites or apps, the vehicle owners deposit their vehi-cles on the operator's platform with their location and availability. The operator then matches these offers with the requests it receives at the same time and thus brings vehicle owners and seekers together. They then meet at the vehicle location, jointly check the condition of the vehicle, exchange the vehicle key, and agree on the time and place of return.

In addition to this matching, the P2P car-sharing operator takes care of a safe and smooth process – not least in order to secure its business model against risks: Vehicle users must register with the operator in advance and prove possession of a valid driving license; vehicle owners, for their part, must provide a roadworthy, registered, cleaned, and fully fueled vehicle. Special installations or modifications to the vehicle are not required. The operator ensures the necessary insurance coverage of the car in car-sharing operation, handles the payment process on its platform, and ensures the most reliable network of sharing partners possible through ranking and appropriate filtering of risk groups among vehicle users and owners.

P2P car sharing is typically billed by the day, with the vehicle owner setting the fee himself and thus being able to control the demand for his car. When handling the payment process, the operator retains a – quite significant – share of the driving fee as a profit margin for himself, usually 25 percent. For the vehicle owner, the temporary provision of his private vehicle represents an opportunity to significantly reduce or even com-pletely neutralize his running costs. A particularly attractive use case is the temporary use of one's own car in situations where high parking fees would otherwise be charged – for example, at an airport parking lot dur-ing a flight. The remaining risk is repair costs due to breakdowns, which are often caused by increased vehicle wear and tear and, unlike accident-related repair costs, are not covered by the operator's insurance.

The business model of P2P car sharing thus functions analogously to the short-term rental of private apartments via platforms such as AirBnB or FeWo-direkt. Also, just like there, another business model is emerging in the shadow of the actual platforms: The initially mainly private providers, who want to reduce their costs by temporarily letting their vehicle to others, are joined on the platforms by more and more providers who acquire – mostly used – vehicles as cheaply as possible and for the sole purpose of offering them in P2P car sharing and thereby making a profit without having to register as a commercial car rental company.

Compared to the B2B offerings mentioned above, however, P2P car sharing is still a clear niche. Most providers such as Getaround in the USA, Easy Car Club or Hiyacar in England, SnappCar in the Netherlands, or iCarsclub in Singapore only offer their services nationally or even locally. Only a few operators are internationally active, such as the American company Turo (formerly RelayRides) or the French company Drivy.

5.2.1.3 Corporate Car Sharing

More and more companies are taking advantage of the benefits of car sharing and offering their employees the use of a corporate car-sharing service instead of an executive car and an employee fleet. The company provides a vehicle pool whose vehicles can be used by employees for business or private trips. Access to the vehicles and the invoicing of journeys are based on the respective company agreements, depending on the type of journey, vehicle class, and the employee's position in the company.

Corporate car sharing has clear advantages over public car sharing: For one thing, all potential users are known as company employees and can thus be held accountable if necessary; careful and disciplined handling of the vehicles is thus far more likely. In addition, it is relatively easy to integrate electric vehicles into corporate car-sharing services, as most journeys start or end at a company location, where a central charging infrastructure can then be provided.

The easiest way for companies to implement corporate car sharing is to outsource it completely to one of the established car-sharing providers. This provider then provides the corresponding user software, handles booking and billing, and takes care of the procurement and overall management of the vehicles. Those who already have a vehicle pool and would like to manage it themselves can implement their corporate car-sharing offer by using one of the many software platforms available on the market.

5.2.1.4   Non-Profit Car Sharing

As early as the 1980s, there were privately organized clubs whose members shared one or more cars for financial or ecological reasons. Similar to the cooperative financing of jointly used buildings or machines in agriculture, the costs of acquisition and operation were apportioned to the members in these clubs according to a usage-dependent key. Rights and obligations were governed by an agreement to be signed by all members. Such cooperative models are particularly attractive because the members are oriented toward the common benefit – for example, they adhere to agreements on vehicle takeover and return, react flexibly to the needs of other members, or pay attention to maintaining the value of the jointly used vehicles. Strictly speaking, the family car shared by parents and children could be described as the simplest form of *nonprofit car sharing.*

This form of privately organized nonprofit car sharing is now being replaced in many places by municipally operated services. In order to improve public transport connections, for example, and thus make it as attractive as possible for their citizens to do without their own cars, many cities operate their own car-sharing platforms at cost price, without any expectation of profit, and are thus able to offer low-cost services that are attractive to citizens.

5.2.2   Acceptance of Car Sharing

So, for whom is car sharing actually an alternative to owning a car? The willingness to completely forego owning a vehicle and instead always rely on being able to borrow a shared car at short notice when needed varies greatly from person to person, even under comparable boundary conditions. Personal preferences and habits play just as much a role here as individual necessities. Also, the decision is at least as much emotional as rational. The decision-making process is guided by three questions: How well do available sharing offers correspond to the personal needs and preferences of the user? What "pain", not only financial, is associated with owning and using a vehicle of one's own? How important is an own vehicle in the respective concrete case, both according to objective and subjective criteria?

5.2.2.1 Match and Attractiveness of the Offer

The first and necessary criterion here is the basic usability of the service for the respective personal mobility needs: Are my place of residence, my workplace, and other destinations that I usually want to travel to within the operating range of the available car-sharing services? The providers primarily cover city centers and neighboring districts close to the city center. However, anyone who wants to use the sharing vehicle to visit friends in the suburbs, for example, may not be able to complete the vehicle booking there but will have to bring the vehicle back to the operating area and park it there. Thus, he pays for the time of stay at the destination as usage time as well. In cities with highly decentralized administration – such as the boroughs in London – it is also quite possible that car sharing is only permitted in individual boroughs and is therefore not available across the board, which dramatically restricts the usability of the service for many.

If the offer basically coincides with personal needs, the question of availability arises: Will I always get a vehicle immediately when I need one? How far do I have to walk to the next available vehicle? Those who want to use sharing vehicles on a permanent basis will generally not accept more than 3 minutes, perhaps even five at peak times. A network of rental stations would have to be extremely dense, which is highly uneconomical and therefore unrealistic, especially in urban areas. Those who want to use car sharing more frequently will therefore resort to free-floating offers. Cities can achieve a hub in availability, performance, and price attractiveness of free-floating car-sharing services by allowing several competitors in the same operating area.

Just as important as the quick pick-up of a vehicle at the departure point is the quick and problem-free completion of the journey at the destination. Reserved parking spaces in inner city areas, for example, contribute enormously to the attractiveness of car sharing. The otherwise possibly lengthy and thus expensive search for a parking space at the arrival point can also be avoided if the provider provides for an *on-the-fly handover* to the next user: In this case, the arriving driver is signaled as soon as he approaches the destination that another user wants to take over the vehicle there right away. The handover can then take place quickly and easily in a short-term parking space or even when the vehicle is parked in a no-parking zone or in a second row.

The next criterion is the vehicles on offer themselves. Do the available vehicles meet my practical requirements in terms of the number of seats needed or size of the load compartment? Are child seats available? Do they also meet my personal tastes, such as my preferences for certain types and brands of vehicles? If particularly attractive vehicles are offered, there is a chance that they will be booked not just out of pure necessity but also for fun, such as a convertible in good weather or a stylish coupé for the evening. Last but not least, the technical and visual condition of the vehicles is also crucial for the acceptance of the sharing offer. The cleanliness of the interior in particular is repeatedly cited by users as a key requirement but is difficult for providers to ensure, especially in the case of free-floating services. Probably the most frequent complaint voiced in user feedback relates to bottles or food packaging left in the vehicle by the previous user.

In addition to these substance-related criteria, the costs of the car sharing offer are also decisive for its attractiveness. Before a motorist decides to give up his own car, he understandably wants to try out alternatives such as car sharing. Here, the price flexibility of the providers plays a decisive role: What do I have to pay before I can use a sharing vehicle for the first time? Fixed costs such as admission fees or regular membership fees spoil the desire to simply try it out. After that, the tariff structure comes into play: Can I charge by the minute? Will the price per minute be cheaper for longer journeys? Having to pay for a whole hour or even a day after a 15-minute ride is absolutely unacceptable for many customers.

Finally, as with all digital services, customer service is critical to the success of car sharing. How quickly, effectively, and friendly help is provided when a question arises during the booking or the journey, or even when a problem occurs, is an important prerequisite for the acceptance of mobility services from the customer's point of view.

5.2.2.2   Inconvenience of Vehicle Ownership

People who never spend long in traffic jams on their way to work and can park free of charge in front of any shop or business understandably have little reason to consider giving up their cars. No matter where in the world, in the countryside and in small towns, the car will still be the predominant means of transport in 10 or 20 years' time.

The situation is quite different in the big cities: There, temporary or permanent entry bans, the deconstruction of roads, and the shortage of public parking spaces are sometimes used quite deliberately to control or reduce inner-city vehicle volumes, while at the same time increasing road tolls, and rising prices for both private and public parking, combined with rigorous penalties for parking violations, are driving up the cost of vehicle ownership. All of this combined is making private vehicle ownership in these cities today less attractive virtually month after month. Vehicle owners eventually look for alternatives out of sheer desperation, and the generation of novice drivers sees their first car, which used to be so important, as less and less desirable.

In addition, car ownership often prevents the use of alternative modes of mobility such as local or long-distance public transport, even if these would be more sensible, faster, or more pleasant than owning a car in a specific case. Those who have already paid for the provision costs such as the purchase price, tax, insurance, or leasing fee for their car are understandably less likely to decide to take public transport to work on the spur of the moment and thus end up paying twice for the journey.

Another disadvantage of vehicle ownership compared to car sharing that has already been mentioned is that, due to the multitude of demands placed on it, owning a vehicle is inevitably always a compromise and thus never optimal for the current purpose of driving. It's usually too big and overpowered for driving to work alone, it could have more space and be more comfortable for driving the family on vacation, and it could be a bit more elegant for going to a restaurant for two. In car sharing, on the other hand, the right vehicle can be chosen for every occasion.

5.2.2.3 Necessity of Vehicle Ownership

However, no matter how suitable the available car-sharing offers are and no matter how many inconveniences are associated with vehicle ownership, everyone attaches a very individual importance or necessity to owning their own car. The following factors, among others, are decisive here:

- The need for special vehicles or vehicle equipment, for example, for craftsmen, large families, or also people with physical limitations or disabilities.
- Many, heavy, or bulky items that should remain in the vehicle such as tools, work documents, samples of goods, sports equipment, or even children's seats.
- The need to have a vehicle available reliably and quickly if necessary, for example, for technical and health emergency services.
- Personal affinity for the vehicle. A true car enthusiast would give up much more than just a convenient way to get around.

A key role in assessing the necessity of vehicle ownership is of course played by the question of whether it is a question of completely doing without a car of one's own or only doing without a second car. The willingness to do the latter is understandably much higher; those who still have quick and uncomplicated access to a vehicle when it matters are much more comfortable with the idea of using mobility services instead of a second car.

In the metropolises, car ownership is becoming more expensive and less attractive month by month. The rental costs for a parking space in the city center have now reached the level of the leasing costs for a mid-range car, and the owner then pays as much for parking as for driving. Even and especially those who can afford it, at some point, simply don't want to go through with it.

5.2.3   Suitable Vehicle Concepts

In addition to the factors discussed in the previous section, the vehicle models used are also a key factor influencing the attractiveness, acceptance, and spread of car-sharing services. The vehicle types, brands, and configurations offered in sharing clearly determine the attractiveness and usability of the service from the point of view of the vehicle user on the one hand and on the other hand also its profitability from the point of view of the operator via acquisition and operating costs as well as the quality of the vehicles. Today, in the early stages of car sharing, standard vehicles with standard equipment are still used almost exclusively for economic reasons. With the increasing spread of car-sharing services, however, it can be assumed that the vehicles used will be much more strongly adapted to the specific requirements not only of the users but also of the operators.

5.2.3.1   Requirements for Car-Sharing Vehicles from the User's Perspective

Those who book a car-sharing service generally do not want to use sharing-specific special models but rather familiar concepts that can be classified according to make and model in the existing automotive world view and thus then also compared and prioritized according to price and performance. Simple vehicle equipment – for example, fabric seats with manual adjustment – is accepted as long as no clear restrictions in functionality or comfort are perceived compared to the models offered in the trade, which would be the case, for example, with the omission of trim, textile coatings, or air conditioning. Car-sharing users therefore want to get the driving experience known or expected from the respective vehicle type and additionally expect the sharing-specific support.

From a holistic perspective, a car-sharing provider must cover the diverse, different mobility needs of potential users as well as possible with its vehicle fleet. In the purchasing situation, in which the customer ultimately makes the decision as to which of the vehicles offered (and thus also which provider) he chooses, he will first look at the objective usability for his purpose of travel but will then certainly also be guided by the subjectively perceived attractiveness of the offers. A sensible market launch strategy therefore consists of initially making available small, maneuverable hatch concepts with little parking space requirement for typical city driving alone or with two people and then larger four-door concepts with up to five seats for further distances and journeys with family or colleagues. Once these basic needs are met, the vehicle fleet can be expanded to include attractive, emotional concepts such as convertibles or coupes, or even more powerful models, which then extend the range upward in terms of price.

5.2.3.2 Requirements for Car-Sharing Vehicles from the Operator's Point of View

When selecting vehicles for its fleet, the provider of car-sharing services is faced with a conflict of objectives. On the one hand, they want to offer attractive vehicles with which they can win over as many customers as possible and also increase their willingness to pay. On the other hand, they are economically dependent on the lowest possible total costs, which are made up of acquisition costs, operating costs, and the residual value upon resale.

Compared to leasing or private ownership, sharing vehicles have many times the mileage due to their use in continuous operation, and they are also subject to comparatively high wear and tear due to the fact that they are sometimes handled with little care. Both factors together mean that they are sold on as used cars after a relatively short time. The economic attractiveness of the offer for the operator is therefore determined not only by the purchase price and the running costs for maintenance and repair but above all by a secure outflow of the used vehicles and the resale price that can be achieved (residual value). For this reason, the operators of car-sharing services generally like to use commercially available and marketable vehicles with a rather low level of equipment and a comparatively high degree of robustness. If maintenance or repair work is required, this selection usually not only affects the costs for working time and spare parts but also via the size of the associated workshop network and the time required to complete the work, during which the vehicle cannot be used in the sharing service and thus does not generate any revenue.

For use in car sharing, special extensions are usually necessary on the vehicles, in particular the installation of the so-called *car-sharing module*, via which the vehicle is reserved by the user, opened, and put into driving mode; the user dialogue is conducted, and the vehicle is locked again at the end of the journey and the fare booking is initiated. Installation and removal of these extensions should be as simple as possible for cost reasons. Examples of this are the following:

- The car sharing module is usually not integrated into the vehicle as an optional extra as part of series production but is only installed subsequently as an accessory. This installation should be as simple as possible, and in particular the module should be easy to remove again in terms of the vehicle's residual value without leaving visible traces such as screw holes or similar in the interior.
- The external identification of the vehicle as a sharing vehicle – e.g., with the operator's lettering – should be carried out as a residue-free removable sticker on a marketable base color. If vehicles are temporarily used for different services (e.g., corporate car sharing during the day and free car sharing at night), magnetic signs that can be quickly replaced instead of stickers can be used for the corresponding identification.
- Parts such as door sill trims, which are subject to particular wear in short-haul sharing operations, should be designed in such a way that they can be replaced without major effort.

A great potential here would be a software integration of the functionality of the car sharing module into the existing system architecture of the vehicle. In this way, the car-sharing module could be represented as part of the vehicle functionality by simple coding without additional hardware and without mechanical intervention in the vehicle and could be removed again by decoding.

The ability to interact *over-the-air* with the vehicles in their own sharing fleet offers operators a number of additional interesting options when designing customer-oriented offers. On the one hand, in the sense of *X on Demand,* individual functions can be activated for the duration of the journey for an additional charge. For example, the use of the sunroof, a parking assistant or a music streaming service in the sharing vehicle, could be booked for a fee and billed at the end of the journey. In the same way, the operator can initially reduce the engine power of its vehicles for new customers or inexperienced drivers and only gradually increase it again after experience and reliability have been proven through the collection of driving data, which can reduce accidents and vehicle misuse and thus costs in the car-sharing business.

5.2.4 Two-Wheel Sharing

In addition to cars, more and more two-wheelers are being offered for sharing, especially in urban areas: bicycles, pedelecs, scooters, motorcycles, and recently also e-scooters, some of them on a massive scale. At first glance, these offers resemble those for cars: The user finds available vehicles via smartphone or at fixed stations, the operator provides the corresponding vehicle fleet, and billing takes place online. However, in terms of their role in the mobility system, the users, and the business model of the operators, two-wheeler sharing offers differ considerably, not only from car sharing but also from each other, depending on the type of two-wheeler involved. Fundamental criteria for whether two-wheelers can be successful as a mobility alternative are, as already described in Sect. 1.2, climatic conditions such as temperature and precipitation as well as topographical conditions such as gradients, but above all also the social acceptance and status that two-wheelers enjoy. In short, where no one rides their own bike, e-bike, or scooter, a corresponding sharing service will not find any friends.

5.2.4.1   Bikesharing

Bikesharing refers to the sharing of bicycles and, more recently, increasingly also electrically assisted pedelecs. While the promotion of carsharing services by local authorities is clearly aimed at reducing the number of private vehicles in the city, bikesharing is of course not associated with the hope that fewer bicycles will be owned. People who already regularly ride their own bikes to work have no reason to switch to a bikesharing service – unless they are afraid that their expensive mountain bike will be stolen or damaged in everyday use. The fact that distances that were previously covered by car are now covered by bike can also be seen as a side effect. The main benefit of bikesharing for cities lies in its enabler function: for many citizens and commuters, the use of public transport only becomes possible or attractive via such a flexible *last-mile solution* for connecting to stations and stops.

Bikesharing is thus primarily a supplement to public transport and is also operated – usually by smaller local companies – in close cooperation with the municipalities. The network of bikesharing stations, as well as fares and payment systems, are integrated into the public transport system: Those who have purchased a ticket for the subway, for example, will ideally find a bicycle for further travel right at the station, the use of which is already included in the fare. Since this positive effect is associated with comparatively low investment and operating costs, bikesharing, unlike car sharing, is not only offered in large metropolises but also in smaller cities and holiday regions.

User Requirements

From the perspective of a bikesharing user, the requirements for such offers are comparatively simple: On the organizational side, the bicycles should be available as widely as possible and be easy to book, open, park, and pay for. In terms of price, use should ideally be integrated into the tariff system of local public transport, and if necessary also into its booking system. As far as the vehicles themselves are concerned, ease of movement (which is in conflict with the required robustness), storage facilities for bags or similar, and, of course, road safety are required. In the case of pedelecs, it should be possible to see the charging status and thus the range before the start of the journey, and there should also be a sufficient number of return stations with charging facilities. In contrast to car sharing, the desire for sporty or emotional vehicles in bikesharing is not discernible in any market.

As with car sharing, a distinction is also made between station-based and free-floating systems for bikesharing. From the user's point of view, the option of being able to park the vehicle directly at the destination and not having to return it to a rental station is naturally much more attractive; for the role described above as a last-mile solution, it is even the only one possible.

Requirements of the Operators

It is the task of the providers to set up sharing stations in the right places within the operating area; to distribute the bicycles of the sharing fleet evenly and in line with demand across the operating area, especially in the case of free-floating offers; and finally to operate the booking and payment systems reliably and in line with customer needs. A decisive aspect for the profitability of this business model is the effort required to maintain the operational readiness of the vehicles offered. While in the case of car sharing the problem lies in the misuse of the vehicle by users – which is not always obvious when the vehicle is returned – in the case of bikesharing open vandalism and theft of parts and vehicles by third parties are also commonplace in many large cities. Time and again, this leads to providers withdrawing completely or partially from markets, such as the recent withdrawal of the Chinese company GoBee from Belgium and France. Against this background, it is also understandable why many operators, despite the advantages of free-floating offers, rely on rental stations in which the bikes are firmly anchored until they can be borrowed.

For the reasons mentioned above, bicycles and pedelecs must also be robust in the first place from the operator's point of view. In order to make it as easy as possible to replace parts when repairs are necessary, while at the same time making theft of bicycles or parts as difficult and as unattractive as possible, bikesharing largely use proprietary components that are, on the one hand, incompatible with conventional bicycles and thus worthless for further use and, on the other hand, are fastened with connecting elements that require special tools that are not commercially available. On the other side of the coin, since the bicycles are thus more or less worthless for private users, resale of used bicycles is also not possible, analogous to car sharing. The vehicles are used until the technical end then have no residual value and are simply scrapped.

Requirements of the Municipalities

At first glance, one would think that municipalities should welcome bikesharing providers with open arms. What is often overlooked is that bikesharing services usually develop into integral components of urban mobility systems, as described above. The focus of municipalities must therefore not only be on the attractiveness of the services in terms of content and price but must also include their operational safety and reliability. Since this is in the public interest, the operators of bikesharing systems are subject to strict requirements on the part of the municipalities, not least with regard to their economic robustness. A sudden unavailability of services, such as the one that occurred in Paris in early 2018 following a change of operator of the Vélib service, which has been in operation since 2007, poses major problems for commuters and citizens who rely on the service, and challenges the municipality to quickly restore its functionality or offer alternatives.

The importance of this control by municipalities over bikesharing systems is also shown by the example of O-Bike. The start-up from Singapore brought thousands of comparatively low-quality sharing bikes to European cities virtually overnight and in some cases without prior approval, which could be booked relatively cheaply and easily used via an app. After a few months, O-Bike filed for bankruptcy and left the cities alone with the disposal of the mostly battered and unroadworthy bikes, which often lay around for months. In order to be at least financially prepared for such cases, many cities now require bikesharing providers to pay a deposit before issuing an operating license.

5.2.4.2   Motor Scooter and Motorcycle Sharing

Taking a plane to a summer holiday and staying mobile between hotel, beach, and sights with a borrowed scooter – a model that has been used for decades. In contrast, the sharing of motorized two-wheelers has only been introduced into everyday mobility in recent years and only slowly. The focus is on offers with large and small electric scooters. The same applies as for cars: if you own your own scooter, you will not borrow another one, at least not at the same location. *Motor scooter-sharing offers* are therefore primarily aimed at people who do not normally drive motor scooters. However, in contrast to bikesharing, which is immediately available to everyone, there are a number of obstacles to the acceptance of scooter-sharing services:

• Driver's license requirement:
  Anyone wishing to drive a borrowed scooter must provide proof of possession of the driving license required for this purpose. For this reason, personal registration is always required for rental and activation, as with car sharing.
• Sense of security:
  Even if a car license is sufficient and available, anyone who does not regularly ride a scooter is not necessarily comfortable with the idea of venturing directly into the traffic of a perhaps even unfamiliar city, completely untrained. With both· car sharing and bikesharing, this inhibition threshold is much lower.
• Limited intermodality:
  Whoever sits on a borrowed scooter then also rides it to the final destination and doesn't change to public transport again – especially not if he sensibly wears a protective jacket and sturdy shoes and perhaps has his own helmet with him. Conversely, the need for precisely these utensils also prevents the spontaneous decision for a scooter as the means of transport of choice for the onward journey.
• Mandatory helmet:
  Some people find having to wear a helmet unpleasant in itself. Putting on a helmet that others have already worn is a downright horrifying

idea for many and a clear criterion for excluding the use of a sharing scooter. For those who have reservations, the only option is to bring their own helmet – which, as already mentioned, can be extremely uncomfortable when riding on the subway, for example. If you have to or want to use the helmet provided by the operator in the storage compartment of the scooter, you have to live with the fact that it just happens to fit properly and that even when worn with a packed cotton hood it remains borderline from a hygienic point of view.

* Zero emissions:
  For many potential users, the noise and exhaust emissions associated with conventional motor scooters are unacceptable, especially in urban areas. Here, only electrically powered scooters come into question.

In view of these hurdles, it becomes clear why scooter sharing is only an option for a relatively small group of users and why it will contribute comparatively little to relieving urban transport systems in the future.

In fact, there is no market for *motorcycle sharing*. They already play only a marginal role as a mobility solution in private ownership; as an element of intermodal mobility, they are even less suitable than scooters for the reasons mentioned above; a broad range of offers does not exist today and is not to be expected in the foreseeable future.

User Requirements

The criteria according to which those interested in scooter sharing then decide for or against an offer largely correspond to those of car sharing: availability, price, and the *"ease of use"* when booking, starting, and completing the journey are the deciding factors as to whether an offer is basically worth considering or not. However, the aspect of safety is much more important for scooters than for cars; users look very closely at the technical condition of the vehicle on offer, especially the function of the brakes and steering or the tire tread. Only then come the requirements for riding comfort (whereby the reservations already mentioned about wearing a helmet are in the foreground) as well as the subjective attractiveness of the available models.

All in all, the barriers listed above show that, from the user's point of view, all scooter sharing offers available today have disadvantages. The reason for this lies exclusively in the characteristics of the vehicles offered. Acceptance could be significantly increased if the scooters used had the following characteristics:

- Electric drive for quiet and emission-free operation with a maximum range of 30–50 kilometers sufficient for urban use
- Can be driven with a normal car license, no moped or motorcycle license required
- Can be ridden without a helmet – and is at least as safe in the event of an accident as a conventional scooter with a helmet

Such a vehicle concept is not yet available from any of the manufacturers but would have the potential to give urban scooter sharing a significant boost and establish it as a sustainable mobility solution.

Requirements of the Operators

Even a comparison of the number of providers makes it clear that the operation of motor scooter sharing is significantly less attractive today than car sharing. One of the main reasons for this is the local weather conditions. On the one hand, booking figures drop drastically when it is cold and raining, and, on the other hand, scooters can only be offered seasonally in many cities due to snowfall. The vehicles then have to be completely collected and stored in winter, which generates additional costs in addition to the lost revenue. On top of this, the disadvantages already mentioned, such as the low level of passive safety and the requirement of a driving license and helmet, severely restrict the potential customer base.

As a result, offering scooters for sharing can only be a successful business model if it represents a sensible alternative within the existing mobility system that can be chosen by the user at any time. The typical use case here is the trip across the city center, where you can get around faster with the borrowed scooter than with a car and do not have to look for a parking space at the end. When designing his offer, the operator must take into account the abovementioned barriers to acceptance. The key here is to select the right vehicles for the specific location and case.

Requirements of the Municipalities

Motor scooter sharing is still not very widespread today and is far less of an integral part of urban mobility concepts than bikesharing. It is questionable whether it can contribute to solving the traffic problems of cities and municipalities in its current form. In this respect, it is hardly surprising that they are rather skeptical about a possible increase or even financial support for scooter sharing. Concrete reasons for this are the following:

- As long as scooters are used instead of cars and as an extension of public transport, road traffic and parking space will be relieved. Like car sharing, however, scooter sharing also has the potential to draw users from public transport onto the roads and thus create additional traffic congestion there.
- The vast majority of scooters in use today are powered by two-stroke engines and thus cause the exhaust and noise emissions typical of them.
- The risk of injury on a scooter is generally comparatively high. With inexperienced riders without protective clothing – as is common in sharing operation – it increases even more significantly.
- Often there are no municipal regulations for the parking of scooters. Parking on sidewalks and public spaces is tolerated in many places, but an increase in scooters, as is to be expected from corresponding sharing offers, would lead to noticeable problems here.

The demand from municipalities for scooter sharing services as well as political approaches for corresponding financial support are correspondingly clearly restrained today.

For an attractive and successful motor scooter sharing offer, the right vehicles are still missing today. A safe, comfortable, and electrically powered scooter that does not require a helmet or protective clothing would make scooter sharing in cities a real mobility alternative.

5.2.4.3   E-Scooter Sharing

E-scooters, the compact pedal scooters with electric drive that are now used worldwide, are certainly not suitable for the daily commute from the suburbs to the workplace in the city center or for other longer distances, but, just like bicycles or pedelecs, they are an effective addition to urban mobility systems. Precisely because they can be ridden on streets and cycle paths and can also be taken along the pavement or an escalator, they are an ideal last-mile solution for inner cities and lend themselves to use in sharing systems. On the other hand, hardly any other means of transport is currently being discussed as controversially as these e-scooters used in sharing, which is primarily due to the often thoughtless or inconsiderate behavior of users when driving and especially when parking the scooters.

Whether or not e-scooter sharing is offered in a city depends primarily on the regulations that apply there – because there are now enough interested providers. In many US and European cities where e-scooters are allowed to be moved on public paths analogous to bicycles (i.e., without operating license, driver's license, or helmet requirement), e-scooter-sharing systems have quickly established themselves. E-scooter-sharing systems are also offered in extensive private environments such as university campuses or company premises. Well-known internationally active operating companies include Bird, Circ, Lime, Scoot, Skip, Spin, and Voi. Some of these companies come from the car-sharing or scooter-sharing business, but there are also independent new players that have grown up offering e-scooters.

The majority of the offers are free-floating systems. After registering via an app, the user locates the e-scooters available in his vicinity via an app and sets them to operational readiness. At the end of the ride, they can be parked anywhere in the operating area, and payment is then made via the app.

With the usual lithium-ion battery, e-scooters typically have a range of 30–45 kilometers. Since in most cases this is not sufficient for the daily mileage required in sharing, the e-scooters have to be charged one or more times a day. This is done via partners of the operating companies, who then also ensure the distribution of the vehicles. Anyone who wants to earn money as a "Bird Charger" or "Lime Juicer" by charging and distributing e-scooters can register online for this purpose and then use a special app to find vehicles with a low charge level, charge them at home or at public charging stations, and then park them at an assigned location and is ultimately remunerated according to their effort. On the one hand, this business model ensures that the vehicles are always charged and optimally distributed in the company's area, and, on the other hand, it saves the operator the investments and official permits required for rental stations and does not require permanent employees.

User Requirements

Availability, price, ease of use, and remaining range are also the primary acceptance criteria for e-scooter sharing. From a technical point of view, the quality of the wheels, the effectiveness of the brakes, easy height adjustment of the handlebars, and sufficient and legally compliant lighting for journeys in the dark are also important.

However, the acceptance of an e-scooter sharing service also depends on how accommodating or rigorous the respective city administration is in regulating its use: In principle, traffic and usage regulations require e-scooters to be driven on the street, but in reality, sidewalks are often used for safety and convenience reasons. Where this is tolerated to a certain extent, or where e-scooters are not allowed to be parked exclusively in specially designated areas, e-scooter sharing will become much easier to establish. In many cities, however, users have now gone too far so that violations of the rules – on the part of both users and operators – are now being rigorously punished.

Requirements of the Operators

Even more so than with bicycles, e-scooters are at risk of vandalism and theft. Since the vehicles are not physically secured – for example, by a bicycle lock or locking at a docking station – they can easily be stolen. From the operator's point of view, it is therefore important, in addition to basic robustness, that the sharing module is integrated as firmly as possible into the vehicle and is therefore difficult to remove or deactivate, as this is the only way to locate the vehicles and thus retrieve them.

In addition, the targeted restriction of the possible group of users is an effective means of reducing the risk of damage due to carelessness. Even if not required by law (in Europe, an e-scooter can already be driven at the age of 14), most operators of e-scooter-sharing systems require a minimum age of 18 for registration or even proof of a driving license – which is not actually required. At the same time, they reserve the right to exclude registered customers from use at any time in the event of misconduct.

One of the decisive factors for the economic viability of the business model is the regulatory framework that the local authorities impose on the operators of e-scooter-sharing systems. On the one hand, this includes requirements such as a restriction on the number of vehicles, regulations for parking vehicles, or deposit payments. On the other hand, cooperation with the abovementioned "independent" service partners is an essential part of the business model. From the operator's point of view, the prerequisite here is that the provision of the service by these service partners (loading and positioning of vehicles) can take place within the framework of an agreement between independent contractual partners and does not constitute a formal employment relationship. Here, too, the municipal legal situation is decisive.

Requirements of the Municipalities

The conflicts to which the rapid implementation of e-scooter sharing services typical of tech start-ups has led, first in American and now also in European cities, demonstrate how important it is to carefully integrate new solutions into existing mobility systems. For example, when several operators in San Francisco, which is actually quite tech-savvy, placed more than a 1000 e-scooters in the city virtually overnight in the spring of 2018 without prior involvement of local authorities, the city reacted with a complete ban after only a few weeks, following increasingly massive complaints from citizens about inconsiderate driving and parking of e-scooters. In a second step, new applications were received, and two providers were selected, who then received approval for the operation of e-scooter sharing systems in the city under restrictive conditions.

In addition to the danger and obstruction of others through reckless driving and parking, the danger to one's own health through lack of experience and overestimation of one's own ability to drive represents a further problem of e-scooter sharing. Typically, after the introduction of such services in a city, injuries skyrocket. The fact that helmets are not compulsory helps acceptance but of course also increases the risk of head injuries in accidents and falls. The data on how dangerous the use of e-scooters actually is and how high the injury rate remains after the introduction phase varies greatly from city to city.

There is no doubt that e-scooter sharing systems are an absolutely sensible element of sustainable urban mobility. In order to effectively counteract the aforementioned excesses and side effects, but at the same time to be able to ensure the continuity of the offer for the citizens, the municipalities would like to retain control over the overall offer of the services provided – as is also the case with bike or motor scooter sharing – for reasons of traffic and regulatory policy. The focus here is on regulations for driving and parking the vehicles as well as punishing corresponding misconduct on the part of the users. However, requirements are also placed on the behavior of the operators, for example, how quickly e-scooters parked in driveways or on wheelchair ramps are removed.

However, both the welfare of the citizens and that of the operators of the sharing services must be kept in mind. If municipal regulation goes too far, e-scooter sharing service providers can withdraw their vehicles from the city just as quickly as they placed them there before. Here, too, the flexibility and fast-moving nature of digital business models that do not require a hardware infrastructure can be seen.

5.3 Ride Sharing Services: Getting Driven Instead of Driving Yourself

For a long time, getting driven was the rule, not the exception, in mobility – initially by carriage, later by train, then by motorized cab or bus. Even at the beginning of the twentieth century, at the dawn of the automobile age, those who were wealthy enough to own their own car did not drive it themselves but had themselves driven by a hired chauffeur – specialists who were equal in rank and standing to a machinist. Just how absurd the idea of driving one's own car to work or anywhere else must have been at the time is demonstrated by the frequently quoted forecast attributed to Kaiser Wilhelm II that the worldwide demand for motor vehicles would never exceed one million, if only for lack of available chauffeurs.

However, when cars became affordable for a much broader group of buyers in the middle of the twentieth century, first in the USA and later in Europe, their private ownership and use slowly became the standard of mobility that it has remained to this day. However, those who could not or did not want to drive their own car in the big city or over longer distances – apart from the exceptional case of the availability of their own chauffeur – were effectively left with only two alternatives until the beginning of the twenty-first century: the use of local and long-distance public transport or the more comfortable, but also considerably more expensive, taxi due to its monopoly position protected by regulation.

Flanked by changes in the legal framework, such as the abolition of the rail monopoly on the operation of long-distance buses in Germany in 2013, as well as technical innovations such as digitalization, a wide variety of new business models for getting driven have been emerging in recent years in precisely this gap, with new providers offering mobility customers a whole range of additional alternatives.

5.3.1 Offers and Business Models

As shown in the previous chapter, the transition from owning a vehicle to car sharing by giving up one's own car not only represents a departure from long-established behavior patterns for vehicle users; it also calls long-established business models into question. It is not only the car trade itself that is feeling the impact of car sharing, especially with the decline in second car sales in conurbations; the demand for accessories and care products is also falling when vehicles are no longer owned. Also, those who previously saw the provision of expensive parking space in city centers as a recipe for long-term economic success and invested in centrally located multi-story car parks will now be closely observing whether and how the number of private cars and thus their clientele decreases as a result of car sharing.

In the next stage of change, the transition from self-driving to getting driven, this effect occurs even more dramatically. On the one hand, the requirements for the vehicles used change dramatically when the paying customer no longer sits in the driver's seat and steers the vehicle but is instead a passenger; on the other hand, this creates completely new service requirements and thus corresponding offers. To the dismay of the long-established players in the automotive industry, who for many years were more concerned with each other in terms of competition, completely new competitors are entering the market and want a share of the big mobility pie. In an unprecedented competitive situation, the established market participants such as vehicle manufacturers, taxi companies, or operators of local and long-distance public transport are suddenly confronted with young, highly innovative and, in some cases, unknown companies from the digital world, which can act quickly and purposefully without a 100 years of brand tradition – but also without the associated legacy costs. For example, among the providers of individual – ride sharing services known today, there is not one that is more than 10 years old:

- Uber, founded in San Francisco in 2009, is considered the inventor of digital ride hailing and a global pioneer in mobility services. Today, Uber is present in over 600 cities on all five continents.
- Also from the USA comes the company Lyft, which was founded in 2012 under the name Zimride as a kind of digital ride sharing agency for long distances. Today, Lyft offers a wide range of mobility services in almost 200 cities worldwide.

- The world's largest mobility provider in terms of number of users is the Chinese company Didi Chuxing, whose services are used by over 550 million registered customers in more than 400 Chinese cities as well as in Australia, Brazil, and Mexico. According to its own figures, Didi Chuxing carried out over 7.4 billion journeys worldwide in 2017.
- Other major international suppliers include India's Ola Cabs, which operates in India, Australia, New Zealand, and England; Indonesia's Go-Jek and Singapore's Grab, which cover Southeast Asia; and Brazil's 99.

What all these providers have in common is a broad product portfolio that is continuously adapted to customer needs: Ride hailing and ride sharing are offered in a wide range of configurations, from motorized rickshaws to limousine services, car, bike, and bicycle sharing, as well as urban transport services such as grocery delivery.

The significant growth potential that continues to be seen in the area of mobility services is shown by a look at the list of quite well-known investors behind these companies. Uber draws on capital from Google and Goldman Sachs, among others. In 2016, General Motors invested 500 million US dollars in Lyft in order to be able to use their experience in the field of autonomous driving, and since 2018 Lyft has also been cooperating with the Canadian automotive supplier Magna. Didi Chuxing is also at the forefront of investors, counting the three major Chinese tech groups Alibaba, Tencent, and Baidu among its backers.

The providers are not only growing through the global scaling of existing business models but are also driving the development of enabler technologies for new mobility offerings. For example, the Uber Advanced Technologies Group is developing solutions for autonomous driving, while the DiDi Research Institute is engaged in applied research in the field of machine learning and image recognition as well as intelligent solutions for traffic management. From the operation of its internationally operating vehicle fleets, DiDi evaluates around 70 terabytes of traffic data every day.

5.3.1.1 Ride Hailing

In principle, the core service of *ride hailing* corresponds to that of a taxi: One or more people call a driver with a vehicle by telephone or hand signal on the street, who then picks them up or gives them a lift and takes them to the agreed destination against payment of a fee. The driver has no destination of his own but is on the road solely for the purpose of transporting passengers for profit. However, ride hailing differs significantly from taxis in two respects: firstly, the operators are independent of associations and regulated licenses, which is why the service and prices can be set freely, and genuine, performance-based competition is created. Secondly, unlike taxis, ride hailing is operated exclusively via a digital platform: Driver and vehicle are always summoned via an app – even if the vehicle is right next to the customer. Transportation is then not provided by trained drivers with licensed vehicles but by independent service providers; payment is not made directly to the driver but cashlessly via the app to the service operator, who in turn then remunerates the driver.

For passengers, ride hailing is thus far more than just a cost-effective alternative to taxis. The underlying digital platform enables a whole range of additional practical features, such as:

- Simple journey request via app by entering the journey destination
- Direct visibility of available drivers/vehicles in the app
- Increased security through specification of driver data, photo, and user ranking
- Forecast of pickup time and arrival time
- Fare estimation before starting the journey
- Simple cashless payment at the end of the journey via app

Uber's success story can also be taken as a prime example of how quickly innovative digital business models can spread in the unregulated space. When Uber started ride hailing operations in 2009, its service formally consisted simply of matching private ride requests and offers to a bilateral agreement and managing payment for it. Although Uber is always perceived as a mobility provider on the part of its customers ("I ride with Uber"), passenger transportation as a service in this business model is formally provided not by Uber but by the driver – and under a private agreement between driver and user. Under current law, Uber charges a fee, like a broker, only for matching supply and demand – and not, for example, for passenger transportation, for which the company also does not consider itself responsible. In this sense, the provider neither employs drivers nor owns its own vehicles, which means that all costs and risks such as loss of earnings due to illness, damage to the vehicle, or claims by passengers in the event of an accident remain with the driver. On the other hand, in order to be able to transport passengers and earn money, drivers only have to prove that they have a valid driver's license and a roadworthy, insured car.

Understandably, ride hailing providers can offer rides much more cheaply than a taxi company on the basis of this business model and thus became an existential threat to the global taxi industry virtually overnight. This was not only due to the fact that passengers migrated to the new, cheaper, and often more quality-conscious competition but above all due to the decline in value of the expensive taxi licenses. In New York City, for example, the price of a taxi license was still around one million dollars in 2013; today, these licenses are already being sold there for under two hundred thousand dollars. It's no surprise, then, that taxi companies, which for years enjoyed a legally protected monopoly situation, are pulling out all the stops to get the new competitors off their backs. This ranges from rallies and demonstrations to lawsuits by interest groups to sometimes questionable and helpless-seeming information campaigns such as that of the London Black Cabs, which inform their passengers on the back of the receipt that the risk of falling victim to violent crime is significantly higher in unlicensed competition. In Cologne, the protest escalated in the spring of 2019 in the open street in the "Uber war" in the form of violence against Uber drivers and their vehicles.

In the meantime, however, legislators have reacted in many places and enacted a series of regulations – varying from city to city – regarding the offer of ride hailing services. In Europe in particular, the business relationship between the service provider and the driver is regulated by law with the aim of ensuring the social security of the drivers, preventing wage dumping and at the same time protecting the competitiveness of the taxi industry. In Germany, for example, only licensed driving service providers with a valid passenger transport permit are allowed to be relayed via ride hailing apps. For employed taxi drivers, Uber and other ride hailing providers are thus suddenly transforming from a threat to a job alternative and a possible path to self-employment. However, lawmakers are also taking action to ensure the safety of passengers, a topic that has been discussed time and again. In the USA, for example, drivers now have to pass a background check, and their cars undergo a technical inspection before they are allowed to offer ride hailing services.

At the same time, licensed taxis continue to benefit from a variety of municipal privileges such as reserved parking spaces especially at high-revenue locations such as train stations and airports or special rights such as dedicated lanes or access permits ("bus and taxi free"). However, the pressure of the newly emerging digital competition has also meant that the scope of services offered by taxi services, which has remained unchanged for a long time, has gained momentum and, at least in some cities, has been expanded to include the option of mobile online booking via a taxi app. Also, another change is noticeable: The newly created competition is leading to more and more taxi drivers paying attention to their driving style, the cleanliness of the vehicle, and their own appearance – much to the delight of the passengers, of course.

However, the municipal administrations are now also listening to the wishes and pressure of ride hailing users, who have become accustomed to the expanded and cheaper mobility offer and no longer want to do without it in the future. On top of that, it is already noticeable that the mobility offer extended by ride hailing has a positive effect on the traffic and parking situation in the city centers.

5.3.1.2 Ride Sharing and Carpooling

What makes the comparison and evaluation of the different ride sharing services significantly more difficult is the inconsistent and ambiguous use of the associated technical terms – not only in everyday language but also through the media. For example, the term *ride sharing* is used both as an umbrella term for individual lift services and for *carpooling*. The following clarification of terms is intended to create clarity here.

Carpooling refers to private ride sharing services in which the vehicle owner – in contrast to ride hailing – takes other people along in his vehicle on a trip that was planned anyway. There can be different reasons for this:

- The full or partial compensation of operating costs by charging a fee, as in the case of the classic car sharing agency
- Sharing operating costs, such as in the case of private carpools to the workplace, where participants take turns driving and sharing rides
- Obtaining other benefits such as permission to use the faster *"carpool lanes"* for multiple occupant vehicles in the USA
- Social responsibility like giving a hitchhiker a ride for free

As with ride hailing, ride requests are made by entering the desired destination, confirming the end of the ride, and payment is made via the associated carpooling app.

Ride sharing, on the other hand, is a special form of ride hailing in which several people with different but relatively well-matched starting points and destinations are transported in the same vehicle. The underlying ride sharing platform ensures that the driver receives well-matched driving orders for the respective route and number of people, which means that he has to make few detours to pick up the passengers and, at the same time, that all available seats in his vehicle are always occupied. For the users, it is because they can share the fare with other passengers and thus travel more cheaply than if they were to travel alone, and, for the operators and drivers, it is because this increases the capacity utilization of the vehicles used, which ultimately means that significantly more money can be earned per minute or per kilometer.

However, whether the driver of a paid ride sharing service really has his own purpose in practice, or whether the ride only serves the commercial transport of passengers, may be significant from a legal point of view but cannot always be clearly clarified in individual cases and is ultimately irrelevant from the point of view of the paying passengers. For the purpose of a clear designation, ride sharing services in which the payment of a fare is in the foreground and the ride is booked and billed via an app are therefore assigned to the *ride hailing* business model *with ride sharing*. If, on the other hand, the focus is on the rider's own journey, then the app-organized provision of rides to other people – whether for a fee or free of charge – is referred to as *carpooling*.

It is understandable that, in contrast to ride hailing services, carpooling services are used much more frequently for long distances between conurbations than for short trips within cities. For the operators, offering ride sharing / carpooling services seems to be far less economically attractive than offering ride hailing services; in many cases, the corresponding apps and platforms are even operated without any economic interests at all – for example, by companies, public authorities, or universities for their own employees or students. Well-known commercial operating companies include BlaBlaCar, Carma, Waze, and carpooling.com. These companies receive a share of the fare for the successful match; in return, they check the reliability of the registered drivers and passengers and provide vehicle and personal insurance for the trips they initiate.

5.3.1.3 Potentials

One aspect of ride sharing that is particularly interesting for cities is its possible integration into public mobility services. If not only passenger cars but also vehicles with more space for passengers and luggage, such as minivans, are used here, the business model comes very close to public transport from the user's point of view. Public on-call buses, which are already used by many municipalities on little-used routes in place of bus routes with rigid timetables, also travel to any point within a certain corridor around the route by call or request in order to pick up or drop off passengers there. The shared taxis used in much of Asia and Africa also represent a – non-digital – form of ride sharing, where the driver adjusts the route based on the destinations of the passengers picked up. If a common platform existed, a highly practical, seamless transition between private ride sharing and public mobility services could be realized from the user's perspective.

Further economic potential lies in the possibility of integrating transport services into ride-sharing services and thus taking letters and parcels on the journey in addition to people. Pick-up and delivery of the freight is arranged via app, integrated into the route planning, confirmed, and paid for. The advantage for the operator is that, depending on the size and type of vehicle, a parcel does not occupy a seat and does not usually complain if it has to take a detour to reach its destination.

Thirdly, the potential frequently discussed in specialist circles should be mentioned here, namely, that, in addition to the fees paid by the users of these services to their operators, there must be further revenue opportunities via additional offers in the area of marketing and sales. This idea is based on the assumption that users of ride sharing services cannot or do not want to occupy themselves with anything else for the usually short duration of their journey and are therefore particularly open to watching advertising clips tailored to them during this period and then buying the products advertised in them online, if possible directly in the vehicle – the account details would already be known and set up via the corresponding app. When the purchase is concluded, the operator then receives a corresponding commission from the provider of the product.

However, apart from the occasional screens with permanent advertising that can be found in Asian taxis and on public transport, this business model has not yet become established anywhere. Such on-board shopping may work on long-haul flights, where there is no or only expensive Internet access, and passengers are therefore unable to use their own smartphones for hours. However, those who are driven through the city and hold their smartphone in their hand anyway because they just used it to call the vehicle they are sitting in use it or enjoy the short rest but certainly do not need an additional online shopping channel. In the digital society, "compulsory" advertising is generally viewed increasingly critically. Just as little, there is no discernible willingness to obtain additional services or reduce the fare by consuming advertising analogous to the familiar pattern of "watch this 20-second clip to the end, after which you can continue to listen to music or watch videos for free". For good reason, no airline, train, or bus line offers reduced tickets for passengers who agree to watch commercials during their flight or journey.

In the inner-city area, car sharing is clearly a transitional solution. A phase that one or the other needs in order to let go of the steering wheel after owning a vehicle. Only then do the real advantages of using mobility services become apparent.

5.3.2 Acceptance of Ride Sharing Services

Anyone who lives in a large city and, due to the ever-increasing maintenance costs and at the same time decreasing total useful life, gives up their own car with a heavy heart and – in order to still be able to drive themselves "like before" – switches to car sharing will very soon discover that driving in the city can also be frustrating in a borrowed vehicle: Where you previously drove around the block for half an hour in your own car looking for a parking space, you now spend just as long looking for a place to park in the car-sharing vehicle – but with the difference that the search now costs 50 cents per minute. At this moment at the latest, the customer then asks himself whether it is really so important to sit behind the wheel himself or whether in the end it would not have been much more pleasant and possibly also cheaper to simply get in, be driven to his destination, and get out again there. Here, it becomes clear why car sharing is a transitional solution in the city – a phase that some people need in order to let go of the steering wheel in a second step after letting go of their own vehicle.

However, for all the pain of owning or borrowing a car, the personal willingness to use ride sharing services instead depends primarily on two factors in the context of the respective individual mobility situation:

* The most important acceptance criterion is availability. This includes the average waiting time resulting from the local coverage with vehicle's and driver's *estimated time of arrival* (ETA); the *service hours*, i.e., on which days of the week and at which times the services are offered; as well as the reliability, i.e., how often no vehicle is available on time after all.
* Only when an acceptable availability is given, the second criterion is the relationship between comfort and price, which is perceived as suitable for the personal situation and the respective purpose of the journey.

5.3.2.1 The Service Gap Between Taxis and Public Transport

The availability of ride sharing services is steadily increasing as a result of the ever new providers in more and more cities and is thus reaching the individual threshold value for more and more mobility customers at which they would be willing to use such services. This threshold is subject to a kind of normal distribution within a mobility system, with only a few people with a very low switching threshold, a majority for whom certain preconditions still have to be created before such a switch can be made, and in turn a smaller proportion of car drivers for whom a switch would not be an option at all or only under the strongest pressure.

As far as freedom of choice in the relationship between price and comfort is concerned, however, the market for mobility services is currently still very clear, as a comparison with other service sectors: if you want to spend a night in a hotel in Munich today, for example, you can choose exactly your personal price-performance ratio for the same type of service (viz., being able to sleep, shower, and have breakfast) on a price scale of between around 50 and over 5000 euros per night. However, if you want to be driven from the main station to the beer garden at the Chinese Tower in the evening, until recently you had to choose between the taxi ride for just under 15 euros and the ride with the subway and bus for 3 euros; there was nothing in between. No wonder that such a gap in supply is an ideal breeding ground for the new digital mobility alternatives from Uber and other providers.

This is because personal requirements and expectations of ride sharing services, as well as the willingness to pay for their use, vary greatly, both within a city and across cities and countries. For example, tuk-tuks, i.e., the three-wheeled motorized rickshaws common in Asia, would hardly be conceivable as a service in New York City; the stretch limousines common there, on the other hand, are at best exotic in major European cities, where in the meantime, however, the bicycle rickshaws widespread in Africa and China are establishing themselves as a flexible and emission-free alternative to taxis, at least in the summer months. The operators of these services can tailor their offerings very precisely to the respective needs: the type of vehicle used, the number of passengers or passenger parties, the functionalities of the booking app, and also the criteria applied in the selection of drivers can be used to determine almost infinitely where the service offered should be placed on the scale between inexpensive and comfortable.

5.3.2.2 Riding in Autonomous Vehicles

Ride sharing services with driverless autonomous vehicles (so-called *robocabs*) are hoping to make a quantum leap in terms of profitability. The underlying calculation is quite simple: autonomous vehicle control eliminates the need for drivers with all their costs, limitations, risks of breakdowns, and accidents, thus reducing operating costs while at the same time making better use of the vehicles and then increasing the bottom-line profit margin. Given this economic potential, service operators such as Uber and Didi Chuxing have been investing billions in the development of fully autonomous Level 5 vehicles for years, alongside car manufacturers and IT companies.

In contrast to the enthusiasm of the providers, however, the comparison between driver-controlled and autonomous vehicles from the point of view of the users of ride sharing services is much more differentiated. Safety and comfort are weighed up against the possible price advantage, whereby emotional aspects are also taken into account in addition to factual ones, as the following points show:

- Anyone who is getting driven not only wants to arrive safely in the sense of unharmed but also to feel this safety during the journey (anyone who has been driven through a large Chinese city on a cycle rickshaw will know the difference). To achieve this, the vehicle should be moved safely, confidently, and as smoothly as possible, and, if necessary, mistakes made by other road users should be compensated for. Especially people who drive a car themselves are often highly sensitive to this when riding along, and, if the expected reaction of the vehicle fails to materialize even for a fraction of a second, they tense up and press the nonexistent brake pedal against the floor panel with all their might, straining into the passenger seat.

Even if the statistics clearly show that the vast majority of traffic accidents are caused by carelessness and thus by human error, the task of moving a vehicle safely can be performed to varying degrees by both human drivers and autonomous vehicle control systems. On the one hand, the spectrum here ranges from the occasional driver who is tired out, distracted by his smartphone or even aggressive, to the confident, professional chauffeur; on the other hand, the systems used for autonomous vehicle control and thus the customer experience when driving a robot taxi can also be of very different quality – from moving forward in a jerky and incomprehensible manner to safely and calmly anticipating the traffic.

Whether a user feels safer driving with a driver or in an autonomous vehicle depends primarily on his or her individual experience. Those who ride along frequently can usually tell how safe a driver is on the road within the first few meters. In addition, even if there are language differences abroad, it is still possible to speak to the driver and ask him to change his driving style, to point out a critical situation that has been overlooked in an emergency or, in the worst case, to stop the journey. It is precisely this theoretical possibility of being able to intervene that contributes greatly to the feeling of safety. In contrast, anyone who climbs into a driverless vehicle is completely at the mercy of its controls. Anticipatory driving, low speeds, and accelerations as well as jerk-free harmonious steering movements promote the feeling of safety here. However, only those who have driven a robot taxi a few times and had a good, safe feeling while doing so will slowly also relax and trust that the autonomous vehicle will not suddenly do things that put them in danger. The common saying in the IT community "To err is human, but to really screw up, you need a computer!" also applies to autonomous driving. Also, the often-used reassurance that people also get into airplanes that are not controlled by humans, but by an autopilot, falls flat when a closer comparison is made: Who would get into a commercial airplane without a pilot and copilot sitting in the cockpit? The decisive factor here is their mere presence, even if the two of them were only there to talk the entire flight.

- In addition to the requirement to arrive unharmed and feel safe on the journey, reliability in the sense of "How high is the risk that the vehicle will not take me to my destination after all, for whatever reason?" is also a relevant aspect for users. What if an unsecured construction site narrows the road on the way? What if black ice has formed on the road? What if the vehicle drives over a nail and a tire blows out? A reasonably experienced driver is trusted to deal with such situations. Besides, is the autonomous vehicle control system also prepared for such cases? In this respect, robotaxis (and any remote control function that may be provided from a control station) still have to earn the trust of their passengers.
- A third aspect of the safety of ride sharing services lies in the subjective feeling that – quite independently of the driving skills – the presence and proximity of the driver triggers in the passenger. Whether this is perceived as positive or negative depends on the one hand on the respective driver but also on the passenger. On the one hand, a driver can make a significant contribution to comfort by, for example, helping passengers get in and out of the car, fetching luggage from the boot, or recommending restaurants to those unfamiliar with the area. Also, just knowing that in case of an emergency there is someone on board who can quickly help or fetch help can contribute to the feeling of safety. On the other hand, passengers may also find the physical proximity of a driver disturbing or even threatening. As described above, ride share operators have responded to reports from their customers of assaults by drivers by raising the selection criteria for their drivers.

The specific situation in which the ride sharing service is used also plays a significant role in answering the question "driver or autonomous control". For example, the demands on the perception of safety are certainly less pronounced during the day in the city traffic of one's own place of residence than at night in an unfamiliar area on a remote country road.

In summary, it can be said that an experienced, reliable, and friendly driver, who also responds to the individual wishes of the passengers, not only conveys safety but also represents a real comfort feature of a ride sharing service, for which certain customers will also be willing to pay a price premium in certain situations. On the other hand, before sitting in a car with someone who makes you feel uncomfortable or unsafe in some way for a short, inexpensive trip through the city, you would much rather be chauffeured to your destination by a robo cab.

5.3.2.3 Safety and Security Related to Ride Sharing Services

One critical aspect with regard to the acceptance of ride sharing services is their safety, both in terms of road safety and, as mentioned in the previous section, protection against attacks (security). Media reports about accidents and especially about acts of violence against passengers – whether well-founded or not – lead to caution and reservations and are only too willingly passed on by competitors such as taxi drivers or companies. Transparency in the sense of a first data-driven insight into the safety situation in ride hailing was provided here by the *US Safety Report 2017/2018*, presented for the first time by Uber for the US market in 2019. According to the report, a total of 3045 cases of sexual violence and 9 assaults resulting in death were reported for 1.3 billion rides provided by Uber in the USA in 2018 – by passengers as well as drivers and by third parties. Surprisingly, the assaults in these cases came from drivers as well as passengers in roughly equal proportions. In addition, 58 people were killed in traffic accidents involving at least one Uber vehicle – again, both drivers and passengers or passersby.

Even though every single case here is of course one case too many, the figures presented also mean, conversely, that 99.9 percent of the trips were completed without any safety incidents. The accident rate per trip as well as per kilometer for rides with Uber is about half of the general values. Moreover, in almost every category in the report, safety levels increased from 2017 to 2018.

5.3.2.4 Ride Sharing Services for Children, the Elderly, and People with Disabilities

When talking about the attractiveness of ride sharing services, the focus is usually on the circumstances under which someone is willing to do without their own vehicle and instead rely on just such services. However, ride sharing services are particularly attractive to people who, because of their age or for physiological reasons, would not be able to drive themselves. Children, for example, but also older people or people with illnesses, injuries, or disabilities depend on someone to drive them to school, to the sports club, to friends, or to the doctor. For many of these user groups, using public transport is too cumbersome or unsafe, and taxis are only available in exceptional cases for cost reasons. New, cheaper but still safe and comfortable mobility offers can bring a significant increase in quality of life here.

It is precisely these groups of people that are often highlighted as a special target group for autonomous driving. However, a closer look reveals that these very people, for the same reasons that they cannot drive themselves, also frequently require a certain degree of supervision and care that varies from case to case. Even if it is just to make sure that nothing has been forgotten in the vehicle when getting out, to reassure someone who has been frightened, or, in more critical cases, to quickly call a contact person, such tasks cannot be performed by a vehicle control system but only by a reliable driver. Despite all the confidence in the underlying technology, what parent would send their 6-year-old child to school alone in a robo-cab in the morning?

Digital ride sharing services can be precisely tailored to local needs in terms of availability, price, and convenience and thus have the potential to close the supply gap between taxis and public transport across the board.

5.3.3 Suitable Vehicle Concepts

Whoever gets a ride today, whether in a taxi, in the hotel's own airport shuttle, or in ride hailing by an Uber driver, in most cases, as with car sharing, gets into a vehicle model like the one offered by the manufacturer as a series model for personal ownership. In the case of ride hailing, this is not surprising, since the rides are, after all, usually provided by independent drivers in their private cars. However, taxis, too, are basically series-produced models that, depending on the country, differ from the respective standard version only in terms of special accessories – such as the taximeter for calculating and displaying the fare, a special paint job or foiling, signals and markings on the outside that identify the vehicle as a taxi and its booking status, and, depending on the manufacturer, minor technical modifications such as reinforced seats or bumpers.

5.3.3.1 Wasted (Horse) Power

In this respect, concepts specifically designed for the transport of passengers would bring a whole range of advantages to the providers of ride sharing services, since the technical requirements for such vehicles differ significantly from the requirements placed on and fulfilled by the vehicles commonly used for self-driving today.

For example, no customer – unless he has booked a race taxi at the Nürburgring as a special form of ride-sharing service – wants to be driven with excessively high longitudinal and lateral acceleration, not even if he is delayed on his way to the airport. However, dynamics and agility are among the primary vehicle characteristics that are demanded in varying degrees by vehicle customers and implemented by manufacturers in their vehicle concepts and in the process make a significant contribution to their manufacturing costs. Today, for example, a Mercedes E 220 D, for a long time the classic taxi vehicle par excellence in Germany, has a drive power of 143 kilowatts, a top speed of 220 kilometers per hour, and a chassis that can easily pull it through any handling course. These features are paid for dearly by the taxi operator when the vehicle is purchased but are not even remotely called upon or needed in passenger service.

However, even beyond dynamics and agility, the overall vehicle concepts of the existing series models are not designed for the passenger but always for the driver. Seating comfort, air conditioning, and entertainment – everything is designed for the driver or the front row of seats.

Many taxi users sit in the front passenger seat because it is often much more comfortable and spacious than the back seat – which should actually be the other way around. However, with the exception of luxury sedans with an extended wheelbase, manufacturers do not yet offer vehicles that are optimized for the customer experience when driving.

In an almost comical way, this discrepancy between supply and demand became clear when a law was passed in India at short notice in 2017, according to which the maximum speed of all vehicles used for commercial passenger transport had to be reduced to 80 kilometers per hour by technical measures. However, not only taxis were affected by the law and the speed reduction but also the shuttles of the large hotel chains and thus hundreds of premium vehicles of the upper and middle classes with their correspondingly powerful six- and eight-cylinder engines. If a manufacturer had had a sedan of the same size in its product portfolio at that time, in which 10,000 euros in manufacturing costs had been saved on dynamics and agility, and then only a fraction of that had been invested in improved driving, seating, and climate comfort as well as on-board entertainment and connectivity for passengers, such a vehicle would have been an absolutely unrivaled offer from the operator's point of view.

5.3.3.2   Requirements from the User's Point of View

If a user has the choice between several ride sharing offers with different characteristics, he will always opt for the offer that best meets his personal price-performance expectations. In any case, the basic service is pure transport to the destination, while the additional service components of ride sharing services can be roughly divided into three categories: ergonomics and seating comfort, entertainment and connectivity, and last but not least safety.

Ride Comfort and Interior Design

How comfortable a passenger finds the ride in a particular vehicle depends on a variety of factors: how effortlessly one can get in and out, how comfortably one sits, how much space and especially legroom one has, how loud or quiet it is in the interior or to which vibrations transmitted by the chassis one is exposed, which may then lead to discomfort. The individual requirement varies here, of course, and also depends on the duration of the journey.

A key component of driving comfort is seat comfort. The criteria for differentiating between vehicle seats are their size, the type of upholstery, the surface material, and the heating and air-conditioning options. For longer journeys in particular, individually adjustable seats are a comfort plus. Ideally, the preferred seat setting can then be saved in the user's personal profile, allowing the seat to be adjusted exactly the same on the next journey. The perceived cleanliness and hygiene of the seat and interior is also an important comfort aspect. Smooth surfaces appear cleaner than fabrics, velour, or other textured materials and still look as good as new even after prolonged use.

Only on longer journeys does the practicality of the interior become relevant. Fold-out tables can be used for working or eating. Suitable storage space allows bags, bottles, smartphones, or similar items to be stored without being forgotten when getting out of the vehicle. When it comes to practicality, particular attention should be paid to local peculiarities; for example, no woman in Asia will want to place her handbag on the floor of the vehicle. Another plus point is the possibility of being able to store drinks for passengers on board and even refrigerate them if necessary.

Special differentiation possibilities in the vehicle interior arise for use in ride sharing. People who ride together also want to be able to sit together and talk. Seats that can be positioned in relation to each other in the manner of a seating area are ideal for this purpose. If, on the other hand, several parties are sitting in one vehicle, suitable facilities for visual and acoustic separation between the passenger parties ensure adequate distance and the desired discretion.

In-Car Connectivity and Entertainment

In cafés, restaurants, and hotels, simple WLAN access with a smartphone has developed from an additional service to a hygiene factor in recent years. At the latest since the abolition of interference liability for the operators of WLAN networks, customers have come to expect free access without time-consuming registration. The same expectation applies to vehicles into which one enters as a mobility customer. Particularly with repeated use, the smartphone should connect all by itself if possible.

While on city journeys most customers tend to want to occupy themselves with their own smartphone, if at all, the behavior on longer journeys is highly different, and at least some of the customers see audio and video entertainment as an attractive additional service. How quickly and directly customer wishes can be responded to in the digital business model here can be well illustrated by an example from ReachNOW, the former US branch of DriveNOW. An analysis of customer feedback there showed that users of the service are most annoyed about two things: when the driver engages them in a conversation when they would rather have their peace and quiet and when a radio station they don't like is playing in the vehicle. As a quick response to this, customers were given the option to indicate whether or not they would like to talk to the driver during the journey and would like to listen to the radio – and if so, which station – when booking the journey via the app. This information is then sent to the driver via his app before the journey begins, so that he can react accordingly. For the passenger, the journey then proceeds exactly as he or she wishes, without having to point this out to the driver personally.

Security

It has already been mentioned that the driver, with his driving style and behavior, is an essential criterion for the passengers' sense of safety. On the vehicle side, the basic functions of passive safety such as a crash-proof body and suitable restraint systems are the main contributors. If passengers are no longer seated in the direction of travel, as is the case in many innovative interior concepts for ride hailing, new technical solutions are required for seats, belts, and airbags.

With additional functions, a further contribution can be made to the sense of security. If necessary, a direct voice connection to the operator or the police can be established via an emergency telephone. It may be important for foreign passengers to be able to communicate in a familiar language. Driverless vehicles in particular should be equipped with intervention options that can be used to stop the vehicle safely in an emergency.

Status

Across all social boundaries, cars also have an emotional, social component beyond their purely functional properties. Different vehicle types and brands consciously or unconsciously send messages about the driver and owner of the vehicle – even if this message is simply "A car doesn't mean much to me". Also, even if among today's 20-year-olds cars have lost a lot of their value as a status symbol compared to their parents' generation, many still have a clear idea of what kind of vehicle they would most like to drive up in on a first date, at a 20-year class reunion, or even on a first visit to their future in-laws.

Here, too, the situation is completely different when it comes to getting driven. Which provider and which vehicle makes the "right" impression on the outside world here? Is Uber really cooler than taxis? Does it always remain rather embarrassing to get out of a stretch limo in front of the club? In a business field that is undergoing as much change as mobility services, there are no long-term statements on this yet. Everything except taxis and public transport is still considered innovative. A differentiated perception will only emerge with the increasing expansion and acceptance of mobility services.

5.3.3.3   Requirements from the Operator's Point of View

Just as with car sharing, the vehicles used in traditional ride sharing services such as taxis or hotel shuttles are initially selected, purchased, and operated by the service provider as operating resources according to purely economic aspects. In the case of the new, digital services such as ride hailing or carpooling, however, the business model and thus the requirements for the vehicles used are completely different; in the end, the service provider only acts as an intermediary here, who in some cases only contributes to the costs of accident and passenger insurance at most. The vehicle, however, belongs to the independent driver, who is also responsible for its purchase and maintenance.

So what requirements does a private individual have for a vehicle with which he or she wishes to transport passengers, at least from time to time, for a fee? This depends primarily on the following criteria:

- Is the vehicle to be used only occasionally or permanently for passenger transport?
- How suitable is the vehicle in general for passenger transport? How spacious and comfortable is the rear row of seats? How easy is the access?
- What are the acquisition and operating costs?
- Is there a reliable yet reasonably priced workshop for the vehicle in the immediate vicinity that can quickly restore the vehicle's availability in the event of damage?
- Has the operator specified any particular class or type of vehicle?
- Must signs or other markings be able to be attached to the vehicle temporarily or permanently?
- For which type of vehicle is the local demand for ride-sharing services the highest? Rather for limousines, station wagons, or minibuses? What is the willingness to pay here?
- Especially if the vehicle is only to be used occasionally for passenger transport: What personal requirements and preferences are placed on it beyond this use?

5.3.3.4   Demand Without Supply

If you look at it closely, both the users and the providers of ride sharing services place a whole bundle of requirements on the vehicles used, which are not met by the standard vehicles available today. The only world-famous exception to this is the *Black Cabs,* which have been used primarily as taxis in London for almost 100 years. Even though their manufacturer has changed several times over this period, this London Taxi with its large and comfortably accessible passenger compartment is the only passenger car concept that was constructed as a purpose design exclusively for commercial passenger transport.

The fact that there are not more offers here is more than surprising in view of the increasing demand for ride sharing services worldwide. Anyone who wants to earn a living as a full-time driver for Uber or Lyft and is looking for a vehicle that would give them a real competitive advantage over other drivers or vehicles thanks to its clear customer orientation is looking in vain on the market. Special equipment or accessory packages that could be used to make series-produced vehicles more attractive, for example, for temporary passenger transport, are also not available anywhere.

However, while in the strictly regulated taxi market with its fixed rules and fixed fares, the motivation to make the vehicles used particularly attractive for customers is rather low (the fare set in the tariff is, after all, independent of this), in the case of digital mobility services such as ride sharing, demand and thus economic success depend heavily on passenger satisfaction as publicly documented in the driver's rating: Entries like "Super comfortable seats, W-LAN access, climate, and entertainment I could set myself" let booking demands grow quickly. At the same time, when transporting several parties in the same vehicle, which is economically attractive for the driver, these could be separated from each other acoustically and visually by flexible elements such as partitions or curtains. Passengers could thus be offered a significantly higher degree of comfort and well-being. What would be needed here are accessories in the areas of ergonomics and interior comfort, entertainment and connectivity, as well as safety, which can be purchased inexpensively and easily installed in the vehicle, but which can also be removed from it again easily and without a trace in the interest of maintaining its residual value.

For lack of alternatives, ride sharing service providers have to use series-produced vehicles whose dynamics and agility they cannot even begin to call up but have to pay dearly for when they purchase the vehicle. Even those who only want to use their vehicle temporarily for passenger transport are searching in vain for appropriate accessories on the market. Vehicles and accessories for ride sharing services with a focus on comfort, entertainment, and connectivity would be unrivalled offerings today.

5.4   Public Mobility

5.4.1   What Does "Public" Mean Here?

In the context of mobility services, the term "public" is initially rather misleading: On the one hand, taxis or ride-hailing services are also accessible to the public, and, on the other hand, "public" buses and trains, as well as ship, train, and airplane lines, are now operated almost everywhere in the world far more frequently by private than by public, i.e., municipal or state, companies. However, the term "public" is used on the one hand to distinguish the mobility service from individual mobility and on the other hand to express the regulation of supply and price by the public sector as well as the public infrastructure required for operation.

5.4.1.1   Public Versus Individual Mobility

The own vehicle, car sharing, taxi, and "normal" ride sharing are assigned to individual mobility. This is characterized by the fact that, regardless of the choice of means of transport, only one party is ever on the move. This can look very different, for example, a single person driving a borrowed electric scooter, a group of colleagues in a taxi, or even the family on holiday in their own car. However, it is always characterized by social togetherness (one is "among oneself") as well as common and freely selectable departure and destination points. Station-based car, bike, or scooter sharing services also count as individual mobility but represent a borderline case with regard to the latter criterion of free choice of departure and destination.

In contrast, with public mobility, several independent parties with different departure and destination points always travel together in the same vehicle. One is therefore "in public" during the journey. The number of people or parties can range from a few passengers in an almost empty regional bus in the Bavarian Forest, to the last seat on a scheduled flight from New York to Chicago, to a completely overcrowded suburban train in London.

The ride-sharing services already mentioned and the shared taxis that are widespread in Africa, Asia, and South America, among other places, in which several parties always share a ride, also belong to public mobility. Even if these services are operated purely privately, they represent the simplest form of public mobility in their own way.

5.4.1.2   Regulation by Public Authorities

On the other hand, "public" as opposed to "private" in this context means, as already mentioned, that the type and scope of the offer as well as its pricing are regulated by the public authorities in order to ensure the – politically desired – basic supply of the population with mobility through corresponding services. Two core requirements are placed on such a basic mobility supply:

• Focusing the offer on the mobility needs of the majority (e.g., connections between suburbs and city centers) and only in a second step on the singular needs of smaller population groups.
• Affordable mobility, especially for users with lower incomes. From a municipal point of view, private mobility, and in particular the use of one's own car, should only be cheaper than the use of public services in exceptional cases.

In local transport in particular, this requires the integration of different municipal and private services into a coherent overall system in terms of spatial coverage (route network), temporal availability (timetable), and pricing (tariff). This requires not only a constant comparison of mobility supply and demand but also the resolution of the natural conflict between the municipal desire for broad coverage, adequate comfort, and low costs on the one hand and the economic goals of private operators on the other. For this reason, municipal transport operators are responsible for defining route networks, timetables, and tariff structures, for selling tickets and, if necessary, for subsidies or compensation payments to private service providers who would otherwise not be able to operate economically under the given framework conditions.

5.4.1.3   Use of Public Infrastructure

In order to be able to offer mobility services at all, it is not enough to purchase vehicles and then operate them with or without a driver – the infrastructure necessary for operation must also be in place: roads and paths on which the vehicles are moved, stops at which passengers can wait and get on and off, but also facilities for supplying energy to the vehicles used.

While for buses this is limited to the installation of signs or bus shelters at the edge of the carriageway or the expansion of separate lanes, the provision of infrastructure for rail-based means of transport is extremely costly. Rails, overhead lines, bridges and tunnels for trams, rapid transit, and especially underground railways, as well as stations and stops, are an integral part of the civil engineering fabric of a city and as such – even if operated by private companies – are owned by them.

Even if not all municipalities can take over the provision and operation of this infrastructure themselves, control over it remains one of the keys to the overall design of mobility systems and should therefore ideally be in public hands. Only in this way can an optimal infrastructure, used equally by public and private mobility service providers, be guaranteed in the long term.

5.4.2   Local Public Transport

Whether it has grown over centuries, as in old cities, or has been integrated into urban planning from the outset, in metropolitan areas and cities, *local public transport* (LPT) is made up of different mobility services on three levels, which need to be coordinated as optimally as possible. From the first to the third level, the passenger capacity, the average travel speed, and the distance between the stops decrease so that an optimal, demand-oriented area coverage then results.

5.4.2.1   Level 1: High-Speed Rail in the Metropolitan Area

The top level of local public transport in metropolitan areas is formed by high-speed trains, which connect the suburbs in the surrounding areas with the urban subcenters of the metropolitan area via mostly radial main lines. Such suburban trains or S-Bahns transport a large number of passengers quickly over long distances; a long S-Bahn train common in Germany, for example, can transport more than 1600 passengers and reach a top speed of up to 140 kilometers per hour. On weekdays in particular, the high-speed trains bring thousands of commuters from the dormitory towns in the surrounding areas to their inner-city workplaces in the morning and back again in the evening. In many cities, large-scale public transport reaches its capacity limits at these peak times – which is a reason for many commuters to continue driving their own cars to and from the city.

Metropolitan railways usually connect different municipalities and administrative areas, in some cases even several cities, which makes the management and regulation of responsibilities for this much more difficult. A good example of such a polycentric transport system is represented by the Rhine-Ruhr suburban railway, which connects more than 50 independent municipalities around the six major cities of Düsseldorf, Dortmund, Essen, Duisburg, Bochum, and Wuppertal.

At the same time, the level 1 rapid transit networks also ensure the transition to the rail lines of public regional and long-distance transport as well as national and international airlines via the connection of city stations and airports and are thus a prerequisite for smooth cross-city mobility.

5.4.2.2   Level 2: Urban Rapid Transit Systems

On the next level of public mobility, a much more closely meshed network of urban rapid transit systems crisscrosses the urban area. These not only connect the major stations with the individual city districts but above all also the districts themselves. Urban rapid transit systems operate underground as a subway or above ground as an elevated railway.

Densely populated, historically developed metropolises offer little space between buildings for rapid transit lines. Here, underground railways or subways provide a fast, congestion-free option for passenger transport within the urban area that is protected from the weather. Among the first cities to be able to afford the necessary underground tunnel construction from the mid-nineteenth century onward were London, Budapest, and Berlin and in the USA Chicago, Boston, and New York. The transport capacity of underground trains depends on their design and can be adapted to demand via the number of carriages – whereby the maximum transport capacity is limited by the maximum possible train length and thus by the length of the existing stations. There are significant differences from city to city: metro trains in Paris, for example, have about 500 seats and standing room, trains in Germany about 1000, and in New York or Tokyo about 1500. The maximum speed in operation is in the range of 50–80 kilometers per hour. Due to the extremely high investment and implementation costs, only relatively few cities worldwide have a comprehensive metro network to date.

In comparison, elevated railways can be built much more quickly and cheaply. Their tracks are usually located above regular road traffic on massive supporting structures made of steel, cast iron, or concrete and, unlike underground railways, are therefore much more difficult to integrate into the existing urban fabric. Classic elevated railway lines from the end of the nineteenth century, such as those in New York, Chicago, or Paris, not only characterize the appearance of cities; the trains running on them also contribute massively to inner-city noise.

Ropeways are a special form of elevated railway that has been used more and more frequently in recent years. These do not require rails and their complex supporting structure and are particularly suitable where urban areas such as residential districts, rivers, or lakes can neither be tunneled or bridged nor travelled over for urban planning, topographical or financial reasons, or where large differences in height have to be overcome. Ropeways were originally developed for winter sports, and their technology has been continuously improved over the decades. Today, they can also be an economically and, in particular, ecologically sensible solution for local public transport. The decisive conceptual advantage over rail-bound railway systems is the separation of the drive motor from the passenger cabins. This means that the motor does not have to be moved along and supplied with energy during the journey but can be permanently installed and connected directly to the power supply system. Even though ropeways are used in many cities such as Rio de Janeiro, Barcelona, or Koblenz to connect sights and thus primarily for tourism, they are also already being successfully integrated into local public transport. For example, the ropeway used in the Bolivian metropolitan region of La Paz/El Alto since 2014 can transport up to 3000 people per hour in each direction.

In all metropolitan areas, accessibility by urban rapid transit has established itself as an important criterion in the residential market. Whether an urban residential location has direct access to the subway or can only be reached by suburban train or bus has a direct impact on its attractiveness and price.

5.4.2.3   Level 3: Light Rail, Buses, and Boats

The third and lowest level is formed by an even tighter network of light railways and buses, as well as boats in some places, from whose stops any destination in the city area can be reached on foot. Compared to buses, trams have the advantage of being able to travel independently of the current traffic situation on the roads. Although buses – unless they have reserved lanes – are stuck in traffic jams with all the other cars, they are much more flexible than trains. Apart from the installation of stops, they do not require any special infrastructure and can therefore adapt their route at short notice to changing demand or traffic conditions. Passenger capacity can also be adjusted in stages at any time by selecting the size of the vehicle, from minibuses to articulated buses. The same applies to passenger ships integrated into the public transport system.

When it comes to the often urgently needed expansion of public transport services, the respective city is faced with the fundamental decision of whether bus or rail lines should be expanded. Both have advantages and disadvantages:

- Rail connections are independent of the current traffic situation and are therefore fast and suitable for very high passenger numbers due to the size of the vehicle. They function (apart from problems with icing) independently of the weather and are therefore generally reliable. However, due to the complex infrastructure such as rails, tunnels, bridges, and power supply lines, the new construction or expansion of rail connections is extremely cost-intensive and, above all, lengthy and thus inflexible.
- Bus services can be implemented quickly and flexibly, but they generally use the same roads as private cars, thus aggravating congestion and at the same time preventing them from making rapid progress

A significant reduction in the number of private cars in cities, which is often desired from a political point of view, would in many places lead to a complete overload of public transport. Striking the right balance here is the core transport policy task of the respective city governments.

Car ferries are a special case here. Although they are part of local and regional public transport, they also serve to continue the journey with one's own car in places where rivers, lakes, or seas cannot be crossed or passed under with bridges or tunnels. For example, the islands of the eastern Mediterranean close to the coast are connected to each other and to the mainland by a dense network of public car ferries. Car ferries thus lie at the intersection of individual and public mobility.

Urban ropeways are still exotic today. However, since the underlying technology is already mature, their operation is particularly favorable in terms of energy, and they can be integrated into existing mobility systems with relatively little effort; ropeways will play an important role as part of local public transport in the future.

5.4.3   Long-Distance Public Transport

Long-distance public transport, in addition to local public transport, covers connections between cities and conurbations, both at national and international level. In view of the long distances to be covered, long-distance public transport includes long-distance buses, long-distance trains, and aircraft, as well as car ferries between port cities. Pure passenger ships play a role on these routes only in the tourist sector as cruise ships.

The political interest in an affordable basic service and thus an active influence of the public sector on the design of services is – in contrast to local transport – rather low in long-distance transport. As mostly independent companies, airlines and long-distance bus operators organize their route networks, timetables, and tariffs purely according to economic considerations, and regulation here takes place at most by avoiding monopolies under cartel law. Long-distance railway lines are also operated on a commercial basis, but here the respective countries have often secured a lasting influence on the design of services through majority shareholdings, even after privatization, as is the case in Germany with Deutsche Bahn AG.

5.4.4 Regional Public Transport Services

Beyond the urban agglomerations, between local and long-distance public transport, lies regional public transport, which has always been rather neglected in comparison to the latter. Anyone in a small town or rural area who wants to use public transport to get to a neighboring town usually has to rely on regional trains or buses, which run at relatively long intervals. If you want to use a regional bus connection outside peak times, you may well have to wait an hour or more for the next bus. Regional bus routes are also characterized by long distances between stops, often requiring a vehicle of one's own to get there from home. In addition, partial connections often belong to different transport associations, which can lead to the customer not only having to change buses on his journey but also having to buy additional tickets.

Due to the vastness and low population density of rural areas, comprehensive and satisfactory coverage by timetabled public transport is not feasible at reasonable cost. No municipality or regional administration can or wants to afford to maintain bus or rail services just in case someone might want to use them, but which then often run empty or with only a few passengers. For this reason, in contrast to the urban area, the own vehicle, depending on age bicycle, motorcycle, or car, is still the predominant mobility solution here. However, all services that operate on user demand, from taxis to ride hailing and ride sharing, have clear potential in this environment.

5.4.5 Business Models

For a long time in Europe, local and long-distance public transport – similar to the postal service and the telephone network – was understood as a self-evident public service, which was provided by the local and state authorities in return for a fare but without the intention of making a profit. Regardless of the actual demand, every station was served strictly according to the timetable, and the staff employed consisted mainly of civil servants. Another important aspect for the operation of state railways, airways, or shipping lines was – apart from providing the population with mobility – the clearly visible demonstration of technical superiority and prosperity vis-à-vis neighboring countries. A good symbol for the cost understanding of such a state railway is the barrier keeper's house: So that every place could be approached, even very remote level crossings were necessary. The official who had to operate the barrier at such a level crossing perhaps once a day was provided with a small house by the state in addition to the official's salary, in which he could live with his family free of charge. Economic efficiency was obviously not an objective of railway operations at that time.

The beginnings in the USA were quite different: in the still young "land of unlimited opportunity", where first and foremost everyone was responsible for themselves, neither the rapidly growing cities nor the federal states and certainly not the Union saw it as their task to look after the mobility of their citizens. It was private entrepreneurs such as the legendary Cornelius Vanderbilt who, in anticipation of long-term economic success, laid tracks across the country, thereby linking cities and coasts with one another, and then made extraordinarily good money from the boom that ensued. The long-distance buses of the private Greyhound

Lines connect even the smaller cities in the USA to this day. Some of the private companies that were involved in the construction of the New York subway network over 100 years ago still operate it today.

These two examples illustrate the tensions in which public mobility has always moved back and forth: on the one hand, there is municipal or state control, which, with high expenditure and rather ponderous structures, aims to ensure a basic supply of mobility that is affordable for everyone – precisely also in less densely populated areas, where the development and operating costs far exceed the revenues. On the other hand, there are private companies with a clear cost-benefit orientation, continuous improvement of services, and economic efficiency as well as efficient decision-making structures, but which ultimately only transport the passengers with whom they also earn money.

There is no silver bullet for resolving this conflict of objectives between state/municipal and private-sector control. In England, for example, 120 railway lines were merged into four private operators in 1923, and these were then merged into the state-owned British Railways in 1948 – which, however, was then broken up into private individual companies again from 1994 onward. Also in 1994, the merger of Deutsche Bundesbahn and Reichsbahn resulted in Deutsche Bahn AG, a company organized under private law, 100 percent of whose shares are owned by the Federal Republic of Germany – thus ensuring the influence of the public sector on operations. However, the "real" privatization originally planned in the form of a stock market flotation has not yet been carried out, not least due to public pressure, and Deutsche Bahn AG is now running profitably. In contrast, the US long-distance and regional rail operator Amtrak, which was privatized in 1991, is subsidized by the government in order to be able to maintain its service – but this obviously undermines the self-responsible pursuit of profitability in this company and has had a strong negative impact on public opinion about these subsidies. In an overall view, the path of integrating individual, competitive private operators through the public sector into an overall system that is coherent in terms of content and economics seems to be more successful than others, at least in the long term.

5.4.6   Acceptance and Potentials

Whether it is local, regional, or long-distance transport or whether someone chooses public transport instead of their own vehicle or individual mobility services, and if so, which one, depends first and foremost on its availability, the duration of the journey, and the comfort of the journey – in each case, of course, in relation to the fare. However, personal preferences are also a major factor in this decision.

5.4.6.1   Availability and Travel Time

In most cities around the world, people who travel purely within the city have no problem reaching their destination by public transport and are usually even much faster than by car. Appropriate apps show even those unfamiliar with the area and language the optimal connection and assist with payment. Based on this high availability, more and more people in metropolitan areas are doing without their own vehicles. However, there are limitations in medium-sized and small cities, for example, during the evening and night hours, where the low demand often does not justify a continuous offer and flexible alternatives are not available.

In long-distance transport, prices and gross travel times for air and rail connections have now converged on many national routes. Long-distance buses, on the other hand, are significantly slower – but also significantly cheaper. On intercontinental routes, air travel is naturally often the only mobility solution; train or ship connections are the absolute exception here. So if you want to travel from city center to city center by public transport, there are usually plenty of highly attractive options to choose from: From Munich city center to Frankfurt city center, for example, you can get there by train almost every hour in just over 3 hours, which is hardly possible by car, even in the best traffic conditions. From Paris to London, the Eurostar also runs hourly in just under two and a half hours, which is unrivalled; by car, you would have barely left the greater Paris area during rush hour.

However, the situation becomes much less attractive in the surrounding areas of the core cities. If you don't want to travel from Munich city center to Frankfurt city center, but instead want to use public transport from Munich's suburbs to Frankfurt's suburbs, the transfer to and from the station may take almost as long as the actual train journey. The situation is even worse if the starting point or destination is in a rural area; a connection without your own car or taxi is then hardly possible.

What would give a real boost to the use of public transport and thus relieve the roads in conurbations of individual traffic would be a significant improvement in the connection of the surrounding area to the public transport network. People who have to get into their cars to reach the nearest train station will think twice before continuing to their actual destination. At the same time, the communities in the surrounding areas are very reluctant to invest taxpayers' money in improving their public transport system just to improve the traffic situation in the neighboring city – after all, the traffic jams are there and not in their own community. Privately operated ride-hailing or ride-sharing services that are integrated into the public transport fare system and connect the working or residential areas of the surrounding areas to the stations of the S-Bahn and suburban trains would be a clever and practical solution to this dilemma.

Another aspect relevant to the acceptance of local and long-distance public transport is the design of transport associations within which different means of transport from different providers can be used with the same ticket. People who travel within a coherent conurbation do not necessarily want to have to change and buy two or even more different tickets just because they cross city or national borders on their journey. The cross-city or even cross-country design of route and tariff plans, the operation of the corresponding means of transport, and the sale of tickets are a huge organizational and political challenge for the authorities involved, but it represents an important key to the acceptance of public mobility services.

5.4.6.2   Costs

As already mentioned, the basic requirement for public mobility is to ensure affordable basic mobility for everyone. However, for those who already own a car, the use of public transport is of course much less interesting financially. In a cost comparison, the travel costs for airplane, train, or bus are then only offset by the operating costs of one's own vehicle, since the expenses for the acquisition – whether purchase or leasing – have already been incurred. This is precisely what makes the modal mix between public transport and one's own car so uninteresting: The car that has already been paid for is parked in a parking lot – usually for a fee – while its owner has to buy an additional ticket to continue his journey by subway.

In many cases, the use of one's own used small car is still cheaper in the full cost calculation than the use of public mobility services. This financial advantage increases even further if additional people travel in the car without additional costs, who would have to pay individually for long-distance public transport. Group discounts are usually only available for six or ten people or more and are therefore irrelevant for the usual private or professional party size.

In order to make public mobility services financially attractive for owners of their own vehicles as well, the legislator therefore has two basic options: On the one hand, the costs per kilometer or per journey with the own vehicle can be raised, for example, by increasing the fuel tax, by making inner-city parking facilities scarcer or more expensive, and by levying road tolls or a city toll. At the same time, the pricing of public mobility services can be used to reduce their usage costs. With this in mind, some cities such as Monheim am Rhein, Dunkirk in France, or Tallinn in Estonia have in recent years already started to offer free public transport to get more drivers out of their own cars and onto buses and trains. The costs of operation are covered by taxes. Luxembourg will be the first country in the world to offer the use of public buses and trains completely free of charge from 2020.

5.4.6.3 Comfort

Even if the importance of comfort decreases in principle the shorter the distance to be covered, many people feel much more comfortable if they are separated from the public, protected from wind and weather, and transported directly to their destination without having to change their means of transport, comfortably seated in a car – whether in a taxi, by ride hailing, in a borrowed car or in their own car – and are also prepared to pay a premium price for this and, if necessary, even to travel longer.

Travelling by public transport, on the other hand, can be particularly uncomfortable for those travelling with luggage. Carrying bulky and heavy luggage on a train, bus, or tram is not only burdensome (especially if you have to change trains a few times) but in many cases simply not possible. It is precisely in this situation that long-distance buses, taxis, or ride-sharing services can score points with customers, if there is enough space to stow their luggage and the driver gives them a quick hand if necessary.

Lack of comfort is one of the most frequently cited reasons for driving one's own car rather than using public transport. In addition to pure driving comfort, other secondary aspects contribute to the comfort experience for many users, such as the possibility to drive themselves, to present themselves to others with their own car, or simply to leave personal items in it. For many people, having their own car also provides a familiar environment that is personally important to them and that they do not want to do without.

5.4.6.4 Safety and Security

When we talk about security in the context of public transport, in the vast majority of cases, we are referring to the likelihood of becoming a victim of theft or violence as a passenger. Perceived and actual safety depend first and foremost on the respective city, the respective line or district, and the time of day at which the journey is made. In many metropolises, residents have developed a feeling for which city districts and thus which sections of public transport connections are better avoided late at night. Security devices such as emergency telephones and cameras in the cars are advantageous for acceptance, as is the noticeable presence of security personnel, which, however, is associated with high personnel costs on the part of the operator. In recent years, many cities have been able to sustainably improve security in public transport in this way. In the New York subway system, for example, there were 26 homicides in 1990 but only one in 2018.

Much less public attention is paid to accidents in connection with public transport than to assaults and acts of violence. Their use is generally considered safe from a technical point of view, and the risk of being injured or killed is still lowest on planes, followed by trains and buses, and highest in cars. However, what is often ignored in the statistics and in people's own perceptions are the dangers of getting on and off trains at stops and in stations. In the Paris Metro area, for example, there are around 400 accidents a year, some of them fatal. To counteract this, operators generally rely on appropriate communication in the vehicles and stations. A particularly successful and noteworthy example of this is the song "Dumb Ways to Die", which was recorded in 2012 in the Melbourne Metro as a safety warning, in which the dangers in the vicinity of stations and trains were pointed out in an entertaining way and which spread virally worldwide on the Internet shortly afterward.

In order for motorists to switch in a big way from their usual and already paid for own car to public transport, two conditions have to be met: the accessibility of stations and stops from home and work and the possibility to use the public services for free.

6

Social Trends

What Are the Societal Conditions Mobility Is Subject to: How Will They Develop in the Future?

Mobility is a basic human need. Just as we – once the basic requirements have been met – are no longer concerned merely with the intake of nutrients to maintain bodily functions when eating and no longer merely with protection from the wind and weather when living, we now generally place significantly higher demands on mobility than simply somehow getting from A to B (a circumstance to which the automotive industry worldwide owes a large part of its revenue …). As discussed in the first chapter, in addition to geographical, infrastructural, and legal circumstances, societal values and trends also determine to a large extent the individual desires with regard to a mobility system and thus, at the end of the day, the individual decision for one of the concrete mobility alternatives. The fact that in recent years, for example, well-off academics in global metropolises are now using high-priced mountain bikes to get to the office instead of cars as before is not due exclusively to the need for healthy exercise or the attempt to escape traffic jams but simply to the desire to be as hip as their friends and colleagues or to follow the trendsetters in the social media. As in other areas of life, social trends in mobility tend to be heterogeneous and differ, sometimes massively, between different spaces, nationalities, or social classes and groups. The acceptance and status of the mountain bike mentioned above would probably be

J. Weber, *Moving Times*, https://doi.org/10.1007/978-3-658-37733-5_6

significantly lower in a small town in the south of the USA than in the City of London, for example.

If such social trends do not remain short-term phenomena but become permanently changed behavior patterns, they will be reflected in legislation and thus ultimately in the design of the mobility system, at least in democratic countries, at very different speeds from country to country and city to city. Politicians who want to be re-elected keep a very close eye on what the population wants and does not want in terms of mobility. A good example of this is the legal requirements for the use of miniature electric vehicles in Germany: The devices such as e-boards or e-scooters that have come onto the market in recent years initially moved in a legal vacuum, and then in 2017 their operation on public roads and paths was generally banned. However, as they represented a reasonable last-mile connection in urban traffic for many users – whether owned or shared – the Small Electric Vehicles Ordinance, which was drafted by the Federal Ministry of Transport and has since come into force, at least allows e-scooters to be operated in public spaces under certain conditions. However, following massive complaints from citizens about inconsiderate driving and parking, this very law is already under scrutiny again in many places.

As the example of the user of an expensive bicycle, who is primarily concerned with his external image, shows, social trends in connection with mobility are multilayered and are interrelated in ways that reinforce or cancel each other out. Nevertheless, a number of guiding strands can be identified in this complex web, which will therefore be discussed in more detail below: first and foremost, the growing desire for comprehensive sustainability, the complicated relationship between society and the automobile, and the efforts of legislators and administrators to steer mobility in the direction of an overall optimum.

6.1 Megatrend Sustainability

6.1.1 Meaning of Sustainability

Although the term *sustainability* originated in the field of nature, it initially had little to do with nature conservation. At the beginning of the eighteenth century, *sustainable management* in forestry referred to the principle of cutting and then selling a maximum of as much wood in the forest as would grow back in the same period. The aim of this self-restraint of the yield was the long-term preservation of the forest but less in its capacity as a natural area worthy of protection than as a profitable asset of the timber industry. Transferred to the general monetary economy, sustainability according to this interpretation means preserving the existing capital and living only from the interest – an idea that, at first glance, seems miles away from the ecologically oriented interpretation of the term that is common today.

It was not until the end of the twentieth century that the concept of sustainability was transferred from the economic to the macroeconomic level. In the 1987 report "Our Common Future", published by the Brundtland Commission, which in retrospect can certainly be described as groundbreaking, the concept of sustainable development was formulated for the first time, as "meets the needs of the present without compromising the ability of future generations to meet their own needs". Here, in addition to *environmental sustainability*, i.e., the preservation of the environment and its resources directed toward the future, *social sustainability*, which is related to the present and aims at the fair participation of all people involved in the value chain for products and services and thus the exclusion of social ills such as exploitation, child labor, and corruption, also appears. It was only later that *economic sustainability*, i.e., the economic safeguarding of the long-term existence of the companies offering the products or services, was also added to the concept of sustainability as a third pillar. Even though economic success and environmental protection were long regarded as conflicting goals, in the shadow of an initial eco- and social romanticism, the insight had slowly gained acceptance that a company must also be economically stable and thus successful in order to be able to operate ecologically and socially in the long term.

The desire to make life on our planet more sustainable is certainly the social trend that has grown most strongly worldwide in recent years. As far as the private environment is concerned, it relates in principle to all areas of life but very specifically to nutrition and clothing – and last but not least to mobility. Whereas in the past it was more the case that individual groups ate healthily, were concerned about environmental protection, or paid attention to fair trade food production, today lifestyles based on holistic sustainability have developed across the board, especially in the major cities. In addition to the traditionally progressive customer milieus, marketing now considers the *Lifestyle of Health and Sustainability (LOHAS) to be* a rapidly growing and quite affluent buyer milieu whose members value their own physical and mental health, but also holistic sustainability as values in life, like to show them to the outside world, and consciously take them into account in all their purchasing decisions.

However, a closer look in the past also showed that the influence of sustainability on individual decisions is also a question of available financial resources and that this often only becomes relevant when the satisfaction of basic needs is ensured. For those who have to ask themselves on a daily basis how they are going to pay for housing, clothing, and food, the question of sustainability is often inevitably of secondary importance. Particularly in the case of food and clothing, sustainability is still a kind of premium property today, reserved only for special, usually more expensive, brands. As long as the regulatory framework is not adapted, for example, in the sense of promoting sustainability, consumers must be able to afford sustainability.

In recent years, this class reference to sustainability has been increasingly overlaid by a generational reference: today's youth and young adults, the so-called *Generation Z*, have grown up with a high level of health and environmental awareness and are increasingly vocal on the Internet and in the streets about their concerns about the way in which the earth is currently being treated, on which they want to spend the next 80 years. They are therefore demanding real sustainability from politics and business, a fundamental change away from the exploitation and destruction of resources toward their preservation. This is a demand that stirs up fears among many older people about the loss of what they have earned over the years and what they consider important or valuable – be it a fur coat, plastic cups for coffee, firecrackers at New Year's Eve, or a high-performance car. Elections and surveys show that the "young" are getting more and more support from the "old", i.e., from their parents' and grandparents' generation. So it's no longer just about tomorrow's customers and voters. This is one of the reasons why politicians and managers are now listening much more closely to what is being demanded on the Internet and in the streets.

6.1.2 Sustainable Mobility

6.1.2.1 Zero Emissions and Resource Conservation

Mobility, and private cars in particular, have always been at the center of any discussion about sustainable living and economic activity. Today, the following aspects are the focus of critical debate:

- Reducing global warming by avoiding greenhouse gases produced and released during the combustion of fossil fuels in vehicle engines, during the manufacture and disposal of vehicles, or during the production of fuel or electrical energy for propulsion.
- Improving air quality by reducing pollutant emissions from motor vehicles. Here, too, not only the use, but the entire life cycle including the energy supply is considered
- The preservation of fossil fuel resources such as oil, gas, and coal, which have been created over millions of years, through the use of alternative and sustainable forms of energy.
- The reduction of public space required by private vehicles for driving and parking.
- The recycling of end-of-life vehicles with as little residue as possible at the end of their lives, which makes the final disposal of problematic materials in particular in landfills superfluous.

It goes without saying that vehicles placed on the market are expected by their users, the legislator, and the public to comply with current emission standards and those foreseeable in the near future. Also, at least in Europe, manufacturers are also expected to take responsibility for taking back and recycling their end-of-life vehicles. However, the diesel scandal of 2015 has quickly led to widespread mistrust among all these stakeholders as to whether the automotive industry as a whole will always live up to this responsibility. The former trust that was placed in the industry has disappeared. Since Dieselgate, on the one hand, the applicable rules and regulations are being critically scrutinized, and on the other hand, manufacturers are also being monitored much more closely in the pursuit of their obligations – not only by the authorities but also by the public.

Most countries today see electromobility as the means of choice for reducing vehicle emissions and conserving fossil resources. In order to achieve the ambitious political goals here, they are investing huge sums in the development of corresponding vehicles and the necessary charging infrastructure. Vehicle drives using hydrogen as an energy source are also still being closely examined, but they are still much further away from the technical maturity required for implementation than pure BEVs. Regardless of how skeptical one is about electric vehicles: if one wants to stick to the agreed emission targets, there is no alternative to a fleet mix with a high proportion of electrically powered vehicles from today's perspective.

6.1.2.2   End-to-End Consideration

Together with electric vehicles and their claim to sustainability, the public has also adopted a much more comprehensive view of sustainability. After critics of electric mobility had repeatedly pointed out that, for example, the combustion of fossil fuels was virtually shifted from the engine to a possibly inadequately filtered lignite-fired power plant, the assessment of vehicle emissions quite correctly included not only the manufacturing and disposal processes. In the so-called well-to-wheel analysis, the emissions associated with the production and distribution of the drive energy were now also included – even though hardly anyone had previously given serious thought to the fuel supply chain from the oil field to the tank and its effects on nature and society.

In such an end-to-end view of electromobility, the question of how sustainably and emission-free the electrical energy with which the vehicle is powered was actually generated is also highly relevant. Here, too, there are highly divergent perceptions both in politics and among the general public. Whereas in countries such as England, France, or the USA, for example, nuclear energy is primarily used and is classified and accepted by the majority there as clean and sustainable, for the majority of people in Germany or the Scandinavian countries only regenerative energy from hydroelectric, wind, or solar power plants fulfils the requirement of sustainability – which certainly represents a hurdle for the spread of electromobility. Ultimately, the sustainability assessment of energy generation must take into account not only the operation but also the production and disposal of the power plants, for example, the construction of an offshore wind farm or the final storage of spent fuel rods from a nuclear power plant.

In order to be able to objectively assess and compare technical solutions in terms of their impact on nature and people, the overall system must always be considered. This applies to emissions and resource consumption as well as to the working conditions of the process partners involved. Sustainability knows no national borders; if you really want to do business sustainably, you have to look at your supply chains right down to the last link. This is time-consuming, but necessary, and can lead to one or two nasty surprises when inspecting the factories of suppliers in emerging countries. For the mobility sector, for example, this includes the conditions under which lithium or cobalt for vehicle batteries is extracted in countries such as Chile, Zimbabwe, or the Congo, as well as the political, military, and technical risks that are taken today to secure oil supplies.

In the holistic sustainability assessment, life cycle assessments or sustainability audits and certificates provide the necessary orientation – provided they have been prepared conscientiously and objectively (i.e., preferably by an independent and accredited institute) and fully cover the part of the process chain under consideration. In some cases, however, these requirements are not met, and the certificates issued are then ultimately not worth the paper they are printed on.

6.1.2.3   What is Sustainability Worth?

Goodwill alone is not enough to make vehicles or mobility services sustainable; as a rule, sustainability also costs money. Anyone who demands that investments be made in occupational health and safety right up to the end of the supply chain and that wages and social security be raised to a fair level will have no choice but to take this extra performance into account in the price negotiations with their suppliers. Also, in order to meet the ever stricter emission limits in all markets, motor vehicles must be equipped with ever more efficient drives and intelligent systems for exhaust gas purification and thermal management – right up to the completely new development of electric drives. Long before the first vehicle is sold and thus earns money, automobile manufacturers must first raise sums in the multi-digit millions for the additional expenditure required in system development and then later also bear the additional manufacturing costs per vehicle built in production.

However, as far as greenhouse gas emissions, which are the focus of regulation, are concerned, the legislators offer an alternative to this. Since the statutory limits usually apply to the entire vehicle fleet sold by a manufacturer, manufacturers of small and rather low-performance vehicles find it much easier to comply with them; manufacturers of both small and large vehicles can offset the higher emissions of the latter with the low emissions of the former. However, those who only offer sports cars, powerful SUVs, or trucks, for example, have no chance of meeting the legally required limits through technical solutions alone but can close the remaining gap by purchasing emission rights (so-called certificates) – so to speak, entitlement certificates issued by the government for the emission of greenhouse gases. Conversely, manufacturers whose fleet emissions fall short of the limits can sell emission allowances on the market in the amount of the excess. Thus, for pure electric vehicle manufacturers with zero fleet emissions, selling emission allowances up to the legal limit is an important pillar of the business model. Tesla, for example, sells US emission rights, the so-called "zero emission vehicle credits", in the order of several hundred million US dollars each year to companies such as General Motors or Fiat Chrysler. Anyone who then takes these revenues into account in the overall economic assessment of a car manufacturer must then, objectively speaking, also attribute the corresponding emissions to that manufacturer, even if, from a purely technical point of view, they do not come from its own vehicles.

At first glance, this business with emission rights is somewhat disreputable; many people have the feeling that rich drivers and manufacturers of large and less environmentally friendly vehicles are buying their way out of their obligations to reduce emissions. On reflection, however, emissions trading primarily leads to the greatest possible efficiency in climate protection. Emissions are reduced first where it costs the least – which, conversely, means that maximum emission reductions are achieved with the available budget. For the actual goal of slowing down or even stopping the rise in global warming, it ultimately makes no difference at all where exactly the production of greenhouse gases was reduced; for the sake of its health, the earth's atmosphere is only interested in the total amount and not the origin. The situation with social sustainability, on the other hand, is completely different. Here, improvements cannot be measured in milligrams of pollutant emissions, nor are such abstract entities as the earth's atmosphere or nature affected, but rather very concrete individuals and their families, not one of whom is more important than the other. Here it is much more difficult to compare the benefits of different measures. Trading in rights in the same way as emissions trading, i.e., buying and selling "social sustainability credits" in the sense of "nothing done about child labor at the African subcontractor, but invested elsewhere in continued payment of wages in the event of illness", would very quickly come up against moral limits. A quantitative optimization of social sustainability based on key figures does not find acceptance.

However, whereas customers are quite prepared to recognize and pay for sustainability as a special and price-worthy product characteristic in the case of food and clothing, it is hardly possible to translate overfulfilment of emission requirements or the offer of additional sustainability aspects in the case of motor vehicles into an additional price. Imagine two identical vehicles – same mileage, same equipment, and same cost of ownership. The only difference is that the second vehicle has 20 percent fewer emissions than the first thanks to innovative engine technology, and many of its parts are made of recycled materials and verifiably come from sustainable supply chains. What extra price would customers pay for the second vehicle compared to the first? The same applies to public mobility: How much more can a ride on an electric bus cost than on a diesel bus? Also, how much more can a subway ticket cost to compensate for the surcharge for green electricity? For individual mobility services such as car sharing, taxi rides, or ride-hailing alone, a clear price differentiation between sustainable and "normal" offers is conceivable.

The question of the added value of sustainability also arises when purchasing an electric vehicle. Especially for manufacturers who offer the same vehicle model with both an internal combustion engine and an electric motor, the latter is usually the most expensive drive variant. After the initial wave of purchases by the rather price-indifferent e-mobility enthusiasts had passed, many prospective buyers asked themselves why they should actually pay a significant extra price for a vehicle with similar engine performance but significantly less range than, say, the comparable diesel model. Only slowly, with the first personal experiences with electric vehicles, is the realization setting in that for this extra price you not only get local emission freedom but also comprehensive sustainability as well as unprecedented driving comfort. Even the newcomer Tesla mainly emphasizes in its marketing roughly comparable range, comparable prices, and superior acceleration values to the combustion engine competition – but cannot really market sustainability per se as a product characteristic. However, if customers are not willing to pay extra for sustainability beyond the legal requirements, vehicle manufacturers and mobility providers cannot be expected to offer it as a free bonus.

6.1.2.4 Social Justice

The close connection between sustainability and social justice has already been mentioned several times. At the core of this is the question of whether, and if so, who has the right to consume or destroy more resources than their fellow human beings. This applies both locally and globally: Is it permissible to emit more exhaust gases or wastewater per capita in industrialized countries than in developing or newly industrializing countries? Are the successful CEO and his wife allowed to cause as much or even more emissions per capita with the heating of their family home than the family of five in the apartment next door? These examples alone illustrate the political and, in some cases, even ideological dimension of social justice, which means that sustainability debates in public are often conducted less technically and objectively than emotionally and subjectively and with almost religious fervor.

Precisely because cars are not only simple means of transport but also serve, on the basis of their style and monetary value, for personal self-expression and the integration of the owner into the social structure, they have always been the subject of particular attention as far as an evaluation of social justice is concerned. How other people live or what they eat is still rather uninteresting from this point of view, what clothes they wear is eyed more critically, but what kind of car they drive is very high on the scale of relevance. This effect could be experienced very clearly when four people died in a tragic accident in Berlin in autumn 2019 because a car driver had suffered an epileptic seizure at the wheel. The apparently most important circumstance in reporting and public opinion was not how one could have recognized such a seizure and thus avoided the accident but that the vehicle had been an SUV of the Porsche brand. The press and politicians spoke of the deadly "city tanks" and discussed banning them, completely disregarding the fact that the consequences of the accident would probably have been no less serious if the driver had been driving a used compact car.

Such emotional expressions of opinion also boil up especially when it comes to political decisions related to mobility – for example, when debating whether first class should be offered in local and regional public transport as it is in long-distance transport, whether there shouldn't be an upper income limit for the free use of local transport, or where in the city a new multistory car park is most urgently needed. As soon as such questions then also involve private vehicles, aspects such as the size, price, and character of the vehicles immediately play an important role: Should the level of road use or parking charges for a vehicle depend on this? Should the purchase of an electric vehicle be subsidized, even if it belongs to the mid-range or even the luxury class? Should SUVs or sports cars be denied access to city centers on principle?

The evaluation and categorization of vehicles for the purpose of forming personal opinions used to be simple and one-dimensional, as the length, weight, engine power, consumption, and emissions of a vehicle still increased more or less evenly with its price; the "big Benz" was synonymous with a large, expensive, and at the same time less environmentally friendly car. Today, this differentiation is much more difficult: an older small car may emit many times more pollutants and noise than a newer vehicle that weighs twice as much and costs four times as much, or even an electric vehicle, which also poses a much lower safety risk to other road users thanks to appropriate control systems.

Regardless of whether we are talking about mobility, housing, food, or general consumption, at the end of the discussion, there is always the rather philosophical question of whether "taking more than you actually need" can ever be sustainable. However, what does "need" mean here? Whether someone really needs a sports car or an upper-class limousine is just as questionable as whether they need to wear fashionable clothes or jewelry, drink red wine with dinner, decorate their apartment with paintings or sculptures, own a holiday home in the mountains, or fly from Munich to Paris for the weekend. How important these things are for the individual in each case, he must decide for himself according to his own subjective criteria (and yes – these are also allowed to change), while the legislators, depending on their effects and risks for the community, provide the framework that determines how easy or difficult these individual desires are to realize.

The currently noticeable trend towards holistic sustainable living and economic activity will no longer be reversed, but rather perpetuated. Mobility offers that ignore this will have no chance in the future.

6.2   Image of Cars in Society

6.2.1   Who Wants Which Car?

Regardless of whether it is bought personally or leased or chosen as a company car, when choosing a car, objective criteria such as driving performance, fuel consumption, reliability, or safety play a role alongside hard framework conditions such as the available budget or the space required. What distinguishes cars massively from pure investment goods (although both are in comparable price regions), however, is the high proportion of emotional aspects in the choice of model and purchase decision. In particular, personal preferences for brands, body styles, or even engine variants depend heavily on perceived membership of certain social groups and their specific trends and behavioral patterns. In this context, the car in its classic role as a status symbol is becoming increasingly less important today, especially in Generation Z., schoolchildren have long since stopped raving about expensive sports car brands, and in the metropolises young adults are increasingly getting by without a car of their own, let alone a driving license. For this generation, new status symbols such as smartphones, brand-name clothing, or even expensive and innovative bicycles apply.

6.2.1.1   Make and Type

Which vehicles are in vogue varies from region to region and is also subject to sometimes rapid change over time. In the USA, for example, the station wagon, a long estate car with a large load space, was the first choice for family cars for a long time, until it was replaced in this role by the SUV virtually overnight at the beginning of the 1990s. Another example are pick-ups, of which there were at most a few imports in Germany until 2010, while in the USA they have always been among the best-selling vehicle concepts. In the meantime, the pick-up wave has swept across the Atlantic, and in addition to established models from Ford and Nissan, pick-ups from VW and Mercedes are now also being offered and bought here.

Overall, there has been a trend in Europe and the USA in recent years toward ever-larger and more powerful vehicles, which is diametrically opposed to the sustainability discussed above and an increasingly car-averse mood in the metropolises. Large SUVs or SAVs with large-displacement combustion engines are in demand among buyers. Also, the manufacturers, who make their revenue with the customers and not with the regulators, are catering to this demand; the model ranges are not only being renewed but expanded upward. Tesla followed up its Model S with the Model X back in 2015, an SUV that is over five meters long and weighs almost two and a half tons. At the end of 2017, Mercedes launched the X-Class, its first pick-up truck, BMW rounded off its model range upward in 2018 with the 8 Series and the X7, and even brands such as Ferrari and Lamborghini or Rolls Royce and Bentley are expanding their range with SUVs beyond the two hundred thousand Euro price mark. At the 2019 North American International Auto Show (NAIAS) in Detroit, light trucks and SUVs predominate, with Ford showing the F150 pick-up there for the first time with diesel power.

On the one hand, this demand is certainly induced in part by manufacturers' supply policies; after all, the earnings from today's business must be used to pay for the investments in innovations such as electromobility, autonomous driving, digitalization, and new business concepts that will then lead to tomorrow's earnings. At the same time, however, in the past imminent disruption has often been heralded by a particularly strong last flourishing of old models of thought that are popular with many: when, for example, in the second half of the nineteenth century, the newly emerging steamships set out to challenge the established sailing ships for their place as the most important means of transport in international maritime trade; it was precisely these sailing ships that experienced a last, new flowering. Hailed as the greatest ship ever built when she was launched in 1869, the Cutty Sark was not only one of the largest and fastest sailing ships ever built but, more importantly, one of the last. It would not be surprising if a particularly large and powerful SUV were soon to meet a similar fate.

The Chinese market is different. Whereas the development of individual mobility from bicycles to scooters, small cars, mid-size cars, and SUVs with maximum engine power in Western countries has often taken several generations, many Chinese have gone through this process in just a few years. Quite a few Chinese BMW 7 series or Mercedes S-class customers, when asked, say that their previous vehicle was a bicycle. It is likely that the highly innovation-savvy Chinese vehicle customers will now also go through the next evolutionary steps such as the use of electric vehicles or mobility services at the same high speed, overtaking the West in the process.

The large and powerful vehicles that are in such high demand today fulfil the customer's wish, encapsulated in the expression *racing saloon*, to be able to fulfil all personal mobility requirements with one and the same vehicle. The requirement here is to transport a family of five quickly, safely. and comfortably over longer distances but at the same time to show status, presence, and sportiness in city traffic on the way to work or to the restaurant. In the future, however, when cars are no longer permanently owned but, for example, in car sharing, are only selected for the journey at hand or, in ridesharing, are not driven at all, this expensive and always compromising universal approach may give way to the technically and economically optimal vehicle for the respective purpose. The role of the car as a status symbol also changes drastically if the vehicle you use to drive up to the restaurant in the evening is quite obviously not yours at all. It can therefore be assumed that in future it will become increasingly rare for the large vehicle needed for the two or three annual holiday trips with the family to also be the vehicle in which father or mother then drive to the office every day for the rest of the year alone and with only a briefcase as luggage.

6.2.1.2 Motors

Public perception and customer desires have also changed significantly in terms of vehicle engine design. Parameters that used to be important for combustion engines, such as displacement or number and arrangement of cylinders, have become less important as a result of efficiency-related downsizing; the plug-in hybrid super sports car BMW i8, for example, uses a three-cylinder engine as its combustion engine. In view of the development of traffic conditions and speed limits, top speeds of over

150 kilometers per hour now play a rather subordinate role worldwide. The only exceptions are Germany, the only country in the world that still has motorways without speed limits, and drivers of sports cars who occasionally drive them on racetracks or at least enjoy the certainty that they could drive their car significantly faster – if only they were allowed to or if the traffic situation permitted it.

In view of the agreed emission targets, politicians have already decided that the passenger car of the future will be electrically powered, at least in most parts of Europe and Asia, and have introduced corresponding legislation. However, the debate among the general public is far from over. Discussions are taking place in the relevant forums on the Internet – on the one hand, factually about topics such as range and availability of charging options and, on the other hand, also highly emotionally, from one side, if sustainability goals are not taken seriously and, from the other side, if there is a threat to established and cherished lifestyles and values. Who holds which opinion depends on the one hand on personal circumstances such as origin, level of education, place of residence, and profession, but on the other hand it is also predominantly a question of age. All in all, it is the *baby boomers* over the age of fifty who do not want to have "their" car taken away from them.

For some, electric vehicles are not only environmentally friendly and sustainable but also innovative and cool; for others, like smoke-free restaurants, vegan food, or faux fur, they are soulless ideological outgrowths of young intellectual city dwellers that miss the point of people's actual needs and desires. It's now part of the standard repertoire of some mainstream comedians to poke fun at drivers of electric vehicles along these lines, and stickers like "F*** Greta" on cars with powerful internal combustion engines are far from the end of the line in terms of expressing one's opinion on the subject. However, as absurd as it may sound, it is the outstanding performance of electrically powered sports cars, which cannot be achieved by any combustion engine vehicle, that has led many a harsh critic to rethink and ultimately accept electric vehicles. In their eyes, such a car is desirable not *because* it is electric but *although* it is electric. In the medium and long term, the personal experience of the comforts of electric drive that go beyond zero emissions; the technical advances, especially in charging and storage technology; and the emergence of the younger generation as buyers, decision-makers, and voters will lead to widespread acceptance of electromobility.

6.2.1.3   Autonomous Vehicles

Unlike electric propulsion, public opinion on autonomous vehicles is primarily characterized by deep skepticism about the technical feasibility and doubts about the fundamental need. Unlike electric vehicles, these are demanded neither by lawmakers nor by the public but primarily by ride-sharing service operators, which is why autonomous vehicles are seen far less as a threat to the personal status quo and lifestyle, and the public discussion about them is thus also much more objective. In any case, no one has yet seen a bumper sticker with a critically provocative message against autonomous vehicles.

Anyone who drives a car has certainly often been in situations that they cannot imagine how an autonomous vehicle control system, no matter how intelligent, could master. This personal perception, together with the recurring reports of fatal accidents caused by autonomous vehicles and a limited level of knowledge about what autonomous vehicles can actually do and where and how they should drive, leads to a rather skeptical basic attitude. The fact that the driver can temporarily leave the controls to his vehicle in traffic jams, when tired or in other situations, is accepted and welcomed as a sensible comfort and safety feature. However, the fact that a vehicle without a steering wheel can chauffeur its passengers safely through urban roadworks or snow-covered country roads seems neither feasible nor desirable in the eyes of many today. As already discussed in Sect. 4.2.2, fully autonomous vehicles, i.e., Level 5 vehicles without steering wheel and pedals, will only be purchased for personal mobility in very few cases – for example, by senior citizens who no longer drive themselves but would like to remain mobile in their own vehicle. First and foremost, fully autonomous vehicles will perform their service in urban areas as robo-cabs, where they can cushion the technical risks of autonomous vehicle control to the maximum and at the same time have the potential to be perceived positively by the public in three ways: by enabling low-cost mobility that is independent of the time of day, by not requiring public parking spaces, and by contributing to increased safety in urban road traffic.

6.2.2 Public Criticism of the Car

Cars have always been much more than just a means of getting people or things from here to there more easily or quickly. Since the beginning of the automotive age, their outer form has moved people emotionally. Design icons by Düsenberg, Rolls Royce, or Bugatti and, above all, sports cars such as a BMW 507, a Ferrari California, a Jaguar E-Type, a Mercedes Silver Arrow, or a Porsche 911 are still admired today across all social classes as works of art, coveted as "dreams in painted sheet metal" and exhibited in museums. In the case of many other models, there is considerable debate as to whether they are beautiful and desirable or not; there are enthusiastic fans and convinced opponents of almost every brand and every model.

However, attitudes toward cars were also determined from the outset not only by discussions of aesthetics and sportiness; they quickly took on an additional socio-critical dimension. The costs associated with owning a car divided society into wealthy people, who could get around quickly and comfortably (and, depending on the model, with the corresponding sensation), on the one hand, and the rest of society, who could not afford a car, on the other. Over the following decades, cars became affordable for an increasingly broad group of buyers and then quickly became indispensable for the majority of the population, both professionally and privately, and consequently also generally accepted in society.

Especially in Germany and the USA (China followed suit at the beginning of this millennium), large and powerful cars have increasingly developed into status symbols, which society views partly with admiration, partly with envy, but especially in the metropolises also increasingly with incomprehension and open rejection. The question of the social justice of vehicle ownership in the light of sustainability has already been discussed above. Particularly critical here are the SUVs, which continue to be in high demand today and which have become the iconic bogeyman of car opponents, at least in the cities, due to their size-related space requirements, their performance-related emissions, and not least their clearly displayed claim to presence. How this critical attitude toward cars will evolve in the future depends mainly on the following four issues: local exhaust and noise emissions caused by vehicles, road safety for pedestrians and cyclists, the need for public space for driving and parking privately owned vehicles, and the environmental impact and political consequences of meeting the demand for fossil fuels for fuel production.

6.2.2.1 Local Emissions

Since the 1960s, vehicle emissions have come under public discussion in major cities. As people in Los Angeles – much like in Beijing today – could barely see their hands in front of their eyes because of smog, closely interwoven social, regulatory, and technical trends developed worldwide, in particular:

- Social and political groups that publicly point out the harmful effects of car emissions on the environment and health and campaign for their reduction
- Emission legislation with binding limit values and measurement methods for motor vehicles, which did not exist until then
- Technical solutions, at the time catalytic converters for exhaust gas aftertreatment and the unleaded petrol required for this purpose

Since then, public sensitivity to vehicle-related emissions has risen steadily and, in line with society's fundamentally emotional relationship with cars, is not always perceived and expressed only rationally and objectively. For example, many members of the social and political groups that publicly denounced environmental pollution caused by car emissions liked to drive secondhand small cars, in Europe, for example, a Citroen 2CV or Renault R4, as a sign of recognition and out of a personal rejection of conservative status symbols – but at the time these cars had no pollutant-reducing measures whatsoever and were therefore comparatively harmful to the environment. The high level of emotionality in the public debate about vehicle emissions has since intensified further. From the turn of the millennium onward, activists in the USA and Europe began in isolated cases to carry out arson attacks on SUV dealers; by contrast, it was comparatively harmless to display vehicles with particularly high CO_2 emissions as pink "climate pigs" for publicity purposes. The protest has certainly reached its peak in terms of publicity so far in the numerous actions at the IAA 2019 in Frankfurt, which were no longer directed solely against pollutant emissions but against passenger cars and the automotive industry as a whole.

So while the totality of vehicle emissions and their global consequences are increasingly viewed critically, the personal view of one's own car remains for the most part comparatively sober and objective. Anyone who buys or leases a vehicle today expects it to meet the applicable emission limits, just like all other legal requirements. The buyer sees his or her personal contribution to reducing emissions as having been made, leaving the responsibility for this to the manufacturer. No vehicle manufacturer has ever offered special equipment – which is technically quite possible – for one of its models, where the reduction of pollutant emissions would have gone beyond the legal requirements, simply due to a lack of demand. Where there has been a clear change in recent times in terms of willingness to invest, however, is in the increasing proportion of plug-in hybrid variants in public and private vehicle fleets, which are also intended to send a visible signal of sustainability to the outside world and, of course, to take advantage of public subsidies where appropriate.

For quite a while, the public debate on the environmental impact of car emissions focused almost exclusively on CO_2 emissions and their impact on global warming. Here, too, cars were the focus of much more criticism than other sources of CO_2 with comparable effects, such as industry or cattle farming. In Europe, social and later legal pressure to reduce CO_2 emissions led to a massive increase in the share of diesel engines, which emit significantly less CO_2, especially in the area of high-performance vehicles, supported by a corresponding reduction in fuel taxes. Diesels were considered reasonable and, after the introduction of soot filters, low-polluting. Only in the USA was diesel still perceived as the dirty and smelly fuel for trucks, and diesel engines led a shadowy existence in the passenger car sector there for a long time. At the same time, environmental groups in the media repeatedly criticized the fact that the emissions determined in the legally prescribed tests under standardized conditions were significantly lower than the values actually occurring in real vehicle operation. However, there was a broad consensus among the public and politicians that vehicle emissions in real-world operation were essentially dependent on the very different individual driving styles, that these could not be fully reflected in any measurement standard, and that the agreed measurement standards therefore served primarily to ensure the objective comparability of vehicles rather than to determine or predict actual emissions. However, this public opinion, which was quite sympathetic toward the vehicle manufacturers, changed abruptly in 2015 when the diesel scandal became known. The fact that a manufacturer manipulates its vehicles in such a way that they recognize when an emissions test is being carried out and, in this case, temporarily reduce emissions to the legally required level led to the loss of the basic trust in the automotive industry described above, which had been built up over many years, to an unprecedented extent, and by and large irrespective of which of the manufacturers had actually and fraudulently carried out such manipulations. Authorities and legislators around the world are now showing a much tougher hand in their dealings with the automotive industry; the discussion of driving bans for diesel vehicles to meet nitrogen oxide limits in major European cities, which gained momentum in 2018, is certainly an example of this.

6.2.3 Traffic Density and Safety

In addition to the pollution caused by exhaust fumes and noise, passenger cars also contribute to two other core problems of cities: increasing traffic density and the safety of all road users on the roads. It is against this backdrop that the reluctance with which some mayors are reacting to the progress made in the field of electromobility must be understood. No matter how happy they may be about the availability of locally emission-free vehicles, traffic jams and accidents cannot be reduced by new drive solutions. International metropolises in particular are therefore looking for suitable measures that could significantly reduce the number of cars on their roads.

The fact that more and more people worldwide are moving from rural and regional areas to the major metropolises, where they naturally want to be mobile, represents the greatest challenge for their transport policy. Particularly in the fast-growing megacities of Asia, Africa, or Latin America, the necessary development of a rail-based public transport network cannot keep up with the speed of population growth, while road-based local public transport is stuck in increasingly dense traffic. At the same time, thanks to the work they have found there, the inhabitants who have moved in can and finally want to fulfil their long-cherished wish of owning their own car, which then ultimately leads to gridlock. However, even in highly developed cities with well-developed public transport, road traffic is subject to a vicious circle and always settles at the maximum tolerable level: as soon as the traffic situation within a city is improved by measures such as a new subway line or a new relief road, the inhabitants, who previously grudgingly used alternative means of transport due to the traffic situation, switch back to the car – until the old "pain level" is reached again. In order to be able to remove the issue of the traffic situation from the list of top problems for the citizens of the megacities, the city administrations need practical and short-term effective control options here. From today's perspective, one key to success here could be AI-based traffic models, on which a large number of international research institutes are currently working.

However, what is often overlooked when considering the flow of traffic, problems with traffic jams are not only experienced by those who are stuck in them themselves with their vehicle (and of course think that all the other motorists around them are responsible for it) but above all by all other road users. Although it is true that the lower the average speed of vehicles, the fewer fatal accidents there are, the vast majority of all road accidents involving personal injury are caused by carelessness on the part of car drivers at low speeds, i.e., when turning, maneuvering, driving in, and, indeed, typical stop-and-go driving during rush hour with too little distance and a high distraction factor. Municipalities can remedy this situation, on the one hand, by introducing technical measures in vehicles, which may also be mandatory, ranging from camera-based pedestrian warnings and autonomous vehicle control to the implementation of a central traffic management system with predefined routing, and, on the other hand, by transforming today's traffic infrastructure, which is usually designed purely for the optimal flow of cars, into a traffic infrastructure designed for the safety of all road users. The focus here is on the intelligent and safe design of pedestrian and bicycle paths and their intersections with each other and with the lanes for passenger cars.

When planning such a transformation, the needs of the various road users must first and foremost be weighed up against each other and prioritized. In this context, municipal designers are repeatedly confronted with the question of what rights and privileges society today still wants to grant to drivers or owners of cars: Is everyone allowed to choose the route they want on the way to their destination? Must every driver have the freedom to break speed limits and parking bans? Is there a right to public parking spaces? In the debate currently reigniting in Germany about a possible speed limit on motorways, the opponents are clearly more motivated by the feeling that they "don't want to be banned from doing everything that used to be fun" than on the basis of rational arguments against such a speed limit.

European pioneers on the question of how to create more space and more safety for pedestrians and cyclists in cities without bringing car traffic to a standstill are certainly Copenhagen, Helsinki, or Eindhoven.

6.2.3.1 Need for Public Space

The fact is that more and more people in big cities are doing without their own cars, some – especially the younger ones – out of conviction and some out of sheer despair about the traffic situation and costs. However, those who don't own a vehicle see the urban design, which is usually oriented toward car traffic, with different, much more critical eyes: multi-lane roadways, bridges, and parking garages everywhere, as well as parking lots at the side of the road. Also, so it is not surprising that from this ever-growing proportion of city dwellers, there are calls for public spaces to be made available for driving and parking private cars to a much lesser extent and instead to be used as urban living space – as a park, as a playground, or even for shops and apartments. "I don't have a car myself. Why should I accept that a large part of the area of my city is reserved for other people's cars?" is the typical line of argument used here by the responsible mayors and city councilors.

In recent years, citizens' initiatives have been formed all over the world on this subject, following slogans such as "Take Your City Back" or "Take Back The Streets" and calling for the dismantling of the car-oriented city. At first glance, this is diametrically opposed to the desire of car owners, who are still present and usually form the majority, for unhindered access to their destinations and parking facilities in their vicinity. Here, too, an objective consideration leads to much better results than an emotional "The cars must all go" versus "We don't want to go back to the Stone Age". As with all urban design changes, the motto here should be to move forward, not backward. In this sense, many of the redesign projects that have been successfully implemented in the meantime show that, thanks to intelligent, space-saving street network design, the impact on car traffic is nowhere near as severe as assumed.

6.2.4 Image of the Automotive Industry

As far as public perception is concerned, the automotive industry has always had a special position. More than in any other industry, not only customers but also real fans identify with the car brands; on the basis of technology and design, the brand and vehicles are often even attributed a kind of "soul". First and foremost Germany as the car nation par excellence with its world-famous brands such as Audi, BMW, Mercedes, Opel, Porsche, and Volkswagen but also France with the charming vehicles from Citroën, Peugeot, or Renault; Italy with sporty models from Alfa Romeo, Ferrari, or Lamborghini; the USA with its often large and powerful vehicles from Cadillac, Chevrolet, Chrysler, Ford, or Jeep (and of course Tesla as the new star in the US car sky); or Japan with world-famous names such as Honda, Mazda, Nissan, Suzuki, or Toyota – the home countries of the manufacturers participate not only in their economic power but also in their image. Thus, even beyond racing, automotive companies are often associated with a degree of national pride. Visible evidence of this is, for example, the fact that politicians in England, which is extremely car-savvy, repeatedly complain in public about the fact that none of the many and traditional English car brands are now in English ownership or that one of the Turkish prime minister's apparently most important projects is the development of a domestic car brand and the production of an associated people's car.

Together with the economic importance of the automotive industry as an employer, taxpayer, and driver of innovation, this consistently positive public image has for a long time led to a cooperative relationship between car manufacturers and politicians, from which the economy as a whole has benefited to a not inconsiderable extent. As already mentioned above, this situation changed abruptly with the diesel scandal in 2015. In addition to the worldwide legal consequences for manufacturers and suppliers, this has caused a dramatic loss of image and confidence across the board for the entire industry. In addition to its customers, who suddenly had to fear for the registrability and resale value of the vehicles they bought, and the environmental protection organizations, which have naturally always viewed it with the greatest skepticism, broad sections of the public and politicians have also been extremely critical of the automotive industry ever since. Even if only a few companies had triggered the diesel scandal through their misconduct, in the public perception and reporting, an entire industry, which had been more or less trusted with the responsibility for meeting its environmental targets, has massively betrayed this trust through fraud.

However, while in other countries the focus was on dealing with the legal aspects of the fraudulent manipulation and, interestingly, vehicle sales of the affected brands did not suffer any significant decline there, the change in public opinion in Germany led to far-reaching consequences. On the one hand, the working methods between industry and politics that had grown over the years were made transparent and critically questioned, especially the emissions targets agreed in the past. Of course, the more intensive the investigations, the more discrepancies came to light. In evaluating them, little distinction was then made as to whether fraudulent action had been taken here or whether applicable regulations had been deliberately violated – or whether, in the evaluator's opinion, the applicable regulations were not simply inadequate. On the other hand, in Germany in particular, the role and future viability of the automobile, and here in particular of the diesel engine, but also of the automobile industry as such, was called into question in politics, reporting, and public opinion and in a way that was in part downright recklessly destructive, despite all understanding for the deficits and need for action that had come to light. The way in which Germany, the "car nation", weakened its own automotive companies after 2015, which are the envy of the whole world, and their vehicles, which still arouse the highest desires all over the world, has caused more than just incredulous head-shaking, especially among foreign observers (and competitors).

It is clear that, particularly in Germany, the automotive industry also has a key role to play in shaping the mobility transformation; where else are the electric, possibly autonomous, and interconnected vehicles with the intelligent and versatile interiors that will be needed to implement sustainable individual and public mobility in the future going to come from? Even though automakers are rich and fed up in the eyes of many, this transformation will require immense creative, organizational, and financial efforts from them. To weaken them further at this stage not only limits their ability to innovate but also threatens their ability to survive, thereby leaving the business and know-how of the future in the hands of international competitors who enjoy far broader support in their home countries.

6.2.5 Giving Back

As already mentioned in several places, passenger cars and their users in many, especially western, cities are facing growing social reservations based on the question of their sustainability and social justice. This opinion is predominantly held by younger people and is therefore naturally increasing throughout society as a whole. From the point of view of car manufacturers or operators of car-based mobility services – both of which, in the interests of economic sustainability, would like to continue developing, manufacturing, and selling vehicles for cities as well, or use them there and earn money with them, for as long as possible – but also from the point of view of city administrations, which have to ensure the political acceptance of their mobility systems, the question therefore arises as to what measures can be taken to counter the reservations and positively influence public perception in the long term.

A promising approach to solving this problem has emerged under the name of *Giving Back*. In the context of social sustainability, this refers to the trend of voluntarily giving something back to society as a person or organization in return for the use of public resources. However, how can passenger cars offset the need for roads and parking spaces, as well as the local emissions and safety risks they pose? This is an additional challenge for car manufacturers, whose vehicles must no longer only meet the requirements of their owners or drivers or, as described in Sect. 5.3.3, of passengers, but must now also offer benefits to previously uninvolved third parties who neither drive nor ride with them but share the common public space. Vehicle functions that enable parked vehicles to contribute to safety or convenience in the city were listed in Sect. 4.4.2.5, for example, an integrated WLAN hotspot, warning of pedestrians running onto the road, use of the vehicle as a safety cell, or access to the first aid box.

Even more than for private cars, the implementation of such functions lends itself to vehicle fleets, for example, in car sharing. In cities that want to drastically reduce the number of cars in the city center in the long term, they can represent a kind of survival strategy for vehicle manufacturers. If passenger cars are to be banned from core cities on a grand scale, cars that all citizens can benefit from will certainly be the last ones left.

Private passenger cars, more than any other contributors to congestion, lack of space and emissions, are under critical public scrutiny. Letting them give something back to society is a promising way to win back their acceptance.

6.3 Social Acceptance of Mobility Services

From the passenger's point of view, the use of mobility services, in contrast to owning a car, still offers little opportunity for differentiation. On the one hand, this creates fewer targets for social appreciation or rejection; on the other hand, however, the lack of opportunity for individual positioning and staging also represents a weighty argument against the use of public and private mobility services for many.

While, for example, planes, trains, and buses for long-distance transport are in no way inferior to the use of a private car in terms of social acceptance, the acceptance of local public transport is determined by three criteria: the local transport situation and thus the possible time advantage, the quality and quantity of the service, and, to a decisive extent, the local social structures and norms. The acceptance of public transport varies considerably from city to city. In London, for example, subways and public buses are used by everyone, regardless of social status, while in other cities the use of public buses is considered unsafe, unclean, uncomfortable, and underprivileged. In a survey of car drivers in Mexico City, for example, the prevailing opinion was that they would rather spend 2 hours stuck in traffic in their own car than sit on a bus next to people they don't know, even if the bus would get them to their destination much earlier through reserved bus lanes or other measures.

The new mobility service providers have now also entered the competition for user acceptance and favor between public services and taxis. Also, while public perception and opinion of the former have been consolidated over the decades, they have yet to be formed for services such as car sharing, ride hailing, and ride sharing due to a lack of broad and long-term personal experience. Currently, they are still primarily seen as innovative and sustainable and therefore cool and trendy. Anyone who rides through the city on a borrowed e-scooter, is picked up from a restaurant by an Uber driver, or arrives at a business appointment in a car sharing vehicle is at least showing their openness to new things as well as a fundamental environmental awareness. At the same time, critical reporting on exploitation, poor working conditions, and a lack of social benefits has led not only legislators but also the public to take a very close look at how operating companies treat their drivers in the interests of sustainability. The safety of mobility services, as mentioned above, continues to be viewed rather critically by the public.

It is also worth mentioning here that the two trends use of mobility services (and the associated renunciation of one's own car) and electromobility go hand in hand in the public perception – although they are technically and organizationally completely independent of each other. Car sharing or ride sharing works just as well or badly with combustion engines as with electric motors. What connects the two topics is the common mindset of a growing group of users who – whether out of reason and pragmatism, idealism, or enthusiasm for technical innovations – have an affinity for both. Those who use an electric vehicle and are thus prepared to accept the restrictions and risks that still exist today are also more willing to do without what they are used to in other areas and, for example, to use mobility services instead of their own vehicle.

6.4   Regulatory Trends

Whether it is the tightening of emission limits for motor vehicles, the approval of new mobility services and vehicle concepts, or the public promotion of alternative drive systems, when social trends become established and develop into majority opinion or expectations, legislation follows suit – at least in democratic countries – with a certain time lag.

6.4.1   Regulation of Mobility-Related Environmental Pollution

For a long time, the focus of legal requirements for motor vehicles was on the safety of drivers, passengers, and other road users. This has been consistently improved by manufacturers over the past decades, and the number of fatalities and injuries caused by road accidents has fallen continuously as a result. So while safety has developed into a hygiene factor and is simply expected by legislators and vehicle customers, today regulations to reduce vehicle emissions are taking up more and more space.

6.4.1.1   Public Opinion

Whether it's the images of smog-darkened Los Angeles and the dead coniferous forests of Europe in the 1980s or the people who now only leave their homes in Beijing and Delhi wearing face masks, public sensitivity to harmful emissions from motor vehicles always increases by leaps and bounds when their effects are directly perceived. For a long time, emissions of pollutants such as carbon monoxide, nitrogen oxides, soot, and other particles were the focus of legal regulations. In comparison to these concrete effects that can be felt in one's own body, the effect triggered by greenhouse gases such as CO_2 and the resulting global warming were rarely perceived personally; after all, here the effect of the emissions is usually felt far away from their origin. This thus rather abstract threat to the environment and health is certainly a major reason why sensitivity to the concentration of greenhouse gases in the air developed comparatively late and at the same time the corresponding scientific forecasts as well as the limit values derived from them are still contested and rejected by parts of the public, economy, and politics today.

In the recent past, however, public opinion on this issue has changed in many places around the world. The sharp increase in climate-related weather phenomena such as hurricanes, floods, or bushfires make the effects of climate change tangible for the population in a direct and painful way. At the same time, in the rapidly growing cities of China or India, the emission of pollutants is increasing on a daily basis, creating the visible and toxic smog that has become a dramatic health hazard for the local population. Worldwide reporting on these environmental consequences, the increased importance of sustainability and health, and a general skepticism about cars with combustion engines that has grown as a result of the diesel scandal are leading to a noticeably growing openness among the population for stricter emission limits, more rigid measures to comply with them, and corresponding demands on politicians. On days when the smog is particularly thick and acrid, car traffic in Beijing or Delhi is drastically restricted, but even in Munich and Stuttgart, where the corporate headquarters of BMW and Mercedes are located and one would therefore expect a certain affinity with car traffic, there is now open talk at the local political level about possible bans on diesel vehicles entering the city in order to be able to comply with the statutory emission limits. Also, even the question of introducing a speed limit on motorways, which was virtually taboo in Germany for a long time, is now being openly debated there again – not against the background of road safety but of fuel consumption and the associated emission of CO_2.

6.4.1.2 Pollutant Emissions and Air Quality in Cities

The total concentration of harmful exhaust gases and particles in the ambient air is decisive for the health hazards to humans and the environment, regardless of where exactly they originate. For this reason, in addition to the emission limit values for polluters, air quality is regulated by legally prescribed pollutant immission limit values, particularly in cities. For cities in the European Union, for example, the *Air Quality Directive 2008/50/EC* not only sets mandatory limit values for maximum concentrations of nitrogen oxides, sulfur dioxide, benzene, carbon monoxide, lead, and particulate matter; it also regulates the relevant measurement requirements, the design and location of measuring stations, and the reporting of measurement results to the European Commission.

Although cars are not necessarily the largest source of pollutants, unlike other emitters, they can be used comparatively easily and quickly as a lever to achieve the limit values in the short term by means of local measures such as access restrictions, speed limits, or driving bans for vehicles with particularly high pollutant emissions. The pressure on cities to comply with the legal requirements is coming equally from the public and from European and national legislators and is steadily increasing. Since 2018, environmental protection organizations, in Germany above all the *Deutsche Umwelthilfe (German Environmental Aid)*, have repeatedly sued cities for exceeding the EU limit values, forcing them to take fast-acting countermeasures. This is because the mayors of the affected cities are ultimately personally legally responsible for compliance with the limit values. The measures of the immission legislation therefore do not affect individual vehicles or manufacturers but always fleets of vehicles, usually differentiated according to their pollutant class.

Buses in local transport play a particularly important role in mobility-related pollutant emissions in urban areas. Numerous cities around the world have already decided to gradually replace old bus fleets with new vehicles with low-emission diesel engines or electric buses. This not only helps the municipalities to comply with the legally prescribed limits but also increases the attractiveness of local public transport and, on top of that, sends a clearly perceptible signal to the public of a political commitment to sustainability.

In connection with immissions caused by mobility, the gases and vapors produced and perceived in the interior of vehicles should not go unmentioned. The majority of these are the so-called *volatile organic compounds* (VOC), which are emitted, for example, from flame retardants, solvents, and plasticizers contained in plastic parts, as well as from paints or adhesives. VOCs are divided into three categories according to their boiling point (i.e., their "volatility") and can be harmless, have an unpleasant odor, or be toxic. As of today, there is neither an internationally recognized definition of VOCs nor legal limit values. The automobile manufacturers in Europe, the USA, and Japan have issued corresponding guidelines that contain measurement regulations and limit values. For the interior of buses and trains, the available guideline values for residential and office spaces are applied in many countries.

Noise is another component of mobility-related pollution that has also been shown to be hazardous to health. The World Health Organization (WHO) has confirmed in extensive studies that noise is the second greatest threat to people's physical and mental health after air pollution and has published corresponding guidelines. Compared to pollutant emissions, however, the development of regulations and legislation in this area is still at a very early stage. For the EU, for example, the *Environmental Noise Directive, which* was published in 2002, does not provide any concrete guidelines such as limit values but rather instructions for action by the Länder and local authorities, for example, on how they can be networked across the board when planning measures to reduce noise.

6.4.1.3   Vehicle Emissions

In contrast to immission legislation, emission legislation relates to the polluters. For motor vehicles, the limits for the new vehicle fleets of the manufacturers are mandatory and are graduated according to the year of manufacture. Emissions legislation is thus directed toward the future and aims to encourage manufacturers in the long term to develop new technical solutions to improve the consumption and pollutant emissions of their vehicles.

Europe

As far as fuel consumption and the directly related CO_2 emissions are concerned, the legislation of the European Union today is the strictest in an international comparison. Here, it was already agreed in 2013 that from 2021 the average of CO_2 emissions calculated over all newly registered vehicles of a manufacturer may be a maximum of 95 grams per kilometer (which corresponds to a fuel consumption of 3.6 liters of diesel or 4.1 liters of gasoline per 100 kilometers), while at the same time this limit is 121 in the USA, 117 in China, and 105 grams per kilometer in Japan.

In order to comply with the greenhouse gas emission reduction targets agreed in the Paris Agreement in 2015, the target of 95 grams of CO_2 per kilometer was further tightened in 2019. In accordance with a new EU regulation, the limit will be gradually reduced by a further 37.5 percent by 2030, i.e., to a fleet average of 59 grams of CO_2 per kilometer. Since it will not be possible to come close to achieving this target by means of further technical improvements to internal combustion engines alone, the new legislation effectively forces manufacturers to at least partially switch to alternative drive systems, which in the given time frame means switching to electric drives.

A further tightening of European emissions legislation is the conversion of the speed-time profile, the so-called driving cycle, which is prescribed for the standardized measurement of vehicle emissions. The *New European Driving Cycle (NEDC)*, which has been in force since 1992, is made up of different phases of constant acceleration, constant speed, and constant deceleration, which makes it easy to implement accurately on chassis dynamometers but which quite obviously reflects completely unrealistic driving behavior. The exhaust emission values determined for a vehicle on the test bench according to NEDC are therefore sometimes significantly lower than the real values measured on the same vehicle during a road journey. In the wake of the public and political mood following the diesel scandal, a new driving cycle, the *Worldwide Harmonized Light Vehicles Test Procedure (WLTP)*, which is closer to real driving behavior, has therefore been mandatory in the EU since July 2017 for measuring emissions and consumption. This is characterized not only by a more realistic speed-time profile, but it also has a higher average speed and average acceleration compared to the NEDC and takes into account the consumption-increasing effects of installed optional equipment. When measuring consumption and emissions, the WLTP leads to approximately 25 percent higher measured values compared to a measurement with NEDC, which means a further significant tension for the achievement of the already strongly lowered limit values by the vehicle manufacturers.

In addition to the limit values themselves, legislation in the European Union is also becoming stricter with regard to checking compliance with the prescribed limit values. In order to ensure that the real consumption values do not deviate too significantly from the values stated by the manufacturer in connection with the type approval, from 2020 newly registered passenger cars must be equipped by the manufacturer with software known as an *on-board fuel consumption meter (OBFCM)*. The OBFCM is used to measure real-world consumption values: fuel consumption for internal combustion vehicles and electrical energy consumption for electric vehicles. In the case of plug-in hybrids, both are determined, so that it is also possible to show the ratio of electric and combustion engine drive in which the vehicle was operated. The consumption data collected will be stored *on-board* by the OBFCM and then transmitted to the European Commission, although the law currently leaves open exactly how this transmission to the authorities is to be implemented – by reading it out during the general inspection, by a measuring device as part of police checks, or *over-the-air* via a mobile Internet connection.

In addition, some EU countries have already committed themselves to long-term goals. Norway, the Netherlands and Slovenia, for example, want to have completely banned combustion vehicles from the roads by 2030, while France and the UK have set themselves the goal of doing so by 2040. Discussions about the timing of the complete electrification of mobility are also taking place in many other countries, but so far without a firm agreement or date.

USA

Unlike in Europe, the focus of US legislation is not so much on CO_2 emissions linked to fuel consumption but on limiting the emission of pollutants with the aim of keeping the air clean. So while, as explained above, the limits for CO_2 in Europe are the lowest in the world, the CARB legislation contains the most demanding limits worldwide for the emission of pollutants, namely, carbon monoxide, nitrogen oxides, hydrocarbons, and particulate matter.

The USA also differs significantly from the EU in the structure of legislative responsibility. Not least because the air quality in its capital Los Angeles has always been particularly poor compared to other major US cities due to the special climatic conditions, the state of California became the pioneer of emissions legislation in the USA. As early as 1966, California passed the world's first law limiting pollutants in car exhaust; a year later, the *California Air Resources Board (CARB)* was founded. At the federal level, on the other hand, the *Environmental Protection Agency (EPA)*, founded in 1970, is responsible for emissions legislation; in the same year, the Clean Air Act was passed, the first federal law on air pollution control.

Since then, CARB has established itself among these two authorities as the driving force in reducing vehicle emissions. Its regulations for technical measures in the vehicle as well as the limits for consumption and pollutant emissions set by CARB have been used as a template for national legislation in many other countries – and in particular have always been significantly stricter than those of the EPA. Today, states in the USA can decide for themselves which of the two sets of laws they want to use. With Connecticut, Maine, Massachusetts, New Jersey, New York, Rhode Island, and Vermont, the eight so-called CARB states have so far adopted California's legislation, all of them states which, like California itself, usually belong politically to the Democratic rather than the Republican camp.

Since the change of administration at the beginning of 2017, the orientation of the EPA in particular has changed significantly. The Republican administration under President Donald Trump questions the greenhouse effect itself as well as the connection between global warming and the corresponding weather phenomena and attaches a much higher priority to America's economic interests – above all those of the domestic automotive industry and energy suppliers – than to national or even international environmental policy goals. As a clearly visible sign, shortly after the change of government, the USA already withdrew from the Paris climate protection agreement signed with 196 other countries in 2016, just 1 year after it came into force. As a result of this political stance, the US government is currently also endeavoring to dismantle CARB's legislative supremacy and to make the EPA more responsible for consumption and emissions targets, which it can then directly access as a federal authority and mitigate the limits accordingly.

In sum, emissions-related regulations for motor vehicles worldwide will continue to evolve in such a way that internal combustion engines and fossil fuels will move from the rule to the exception in the long term.

6.4.2 Regulation of the Car Population

Quite apart from curbing pollutant and greenhouse gas emissions from motor vehicles, those responsible in city administrations have another task, namely, to keep traffic flowing on their streets and to provide sufficient parking facilities. For a long time, the measures of choice here were the expansion of the urban road network and the construction of multi-story car parks in the center – but this reached its limits early on, especially in Europe's historically grown inner cities, which is why the regulatory focus there shifted from facilitating car traffic to limiting it. Whereas in Paris or London, for example, further elaborate integration of streets and parking garages into the existing building fabric is hardly possible, in Los Angeles, for example, car traffic has been part of urban planning from the very beginning.

In combination with the goal described in the previous chapter of reducing emissions caused by car traffic, limiting or even reducing the total number of cars in the city is therefore one of the most urgent tasks for administrations in most large cities. Municipal measures in this respect can pursue two fundamentally different thrusts: On the one hand, the, rather long-term and expensive, promotion of the availability and attractiveness of alternatives such as local public transport or private mobility services and, on the other hand, the, short-term and comparatively inexpensive, targeted discouragement of the use of one's own car by limiting or increasing the cost of acquisition and use, in the latter case differentiated according to driving and parking.

6.4.2.1 Restrictions on Acquisition and Use

The most important criterion from a political point of view when designing regulations to restrict the purchase or use of private cars is usually their social compatibility, i.e., taking into account, for example, occupational or health-related necessities of vehicle use, or also the political desire not to make driving in the inner city the privilege of wealthy citizens by simply increasing road use and parking charges. Different cities are taking different approaches to this. Three examples show the range of possible measures:

- In London, the *London Congestion Charge of* 10.00 pounds per day (about 12.50 euros) has been levied for entry into the city center between 7 a.m. and 6 p.m. since 2006. The revenue from this goes directly to the operator of the public transport system *Transport for London*. The main criticism of this scheme is that the toll is levied as a flat rate and not in relation to actual road use.

  London leads the world in the cost of parking a car. The price of buying or renting a private parking space in the city center is the same as that paid elsewhere for a spacious condominium. Those who find one of the rare parking spaces at the side of the road pay between 4.00 and 6.80 pounds (4.69 or 7.98 euros), depending on the type of vehicle. A parking ticket costs up to 130.00 pounds (152.58 euro), towing additionally 200 pounds (234.71 euro).
- In Beijing, a number of measures were taken to reduce car traffic in the run-up to the 2008 Summer Olympics. Since then, for example, vehicles are not allowed to drive in the city center on one working day per week that depends on the last number of their registration plate, which

reduces the volume of traffic there by 20 percent. To further limit this, since 2011 the number of newly registered passenger cars has been strictly limited by a quota of the number plates required for them. Anyone wishing to register a vehicle must take part in the *Beijing License Plate Lottery*, which takes place every 2 months, where they then have the chance to win permission to purchase a license plate. The probability of winning is now around 1:2000, so it is no wonder that license plates are sold or rented out on the black market at horrendous prices, despite severe penalties.

The city of Shanghai, on the other hand, takes a much less communist approach. Here, license plates are not raffled off, but auctioned off to the highest bidders. In 2018, the average price achieved for a license plate was around 88,176 yuan (about 11,500 euros).

- The city state of Singapore also rigidly regulates the number of passenger cars on its roads. A certificate costing 50,000 Singapore dollars (about 33,500 euros) is required for registration, and at the same time the annual growth of vehicles in the city has long been limited to 0.25 percent. Since 2017, the total number of vehicles has been kept constant, meaning that only as many new vehicles are registered as old ones are deregistered.

6.4.3 Financial Support

6.4.3.1 Promotion of Electrically Powered Passenger Cars

In addition to direct subsidies at the time of purchase, permanent or time-limited, full or partial exemption from motor vehicle tax, tolls or parking charges, and free charging options, in particular, also represent further forms of financial purchase incentives for vehicle types preferred by the legislator. At the same time, the selective lifting of restrictions applicable to conventional vehicles (such as driving bans in inner cities) offers the possibility of nonfinancial support. Again, concrete examples best illustrate the range of possible measures:

- The prime example of state-subsidized e-mobility is Norway. Here, the otherwise crisp value-added tax of 25 percent is waived on the purchase of an electric vehicle, and there is no vehicle tax or exhaust emissions levies on its operation, at least until 2025. Road use, ferries, and parking in public car parks are significantly cheaper, and free charging stations are available in car parks reserved for electric vehicles. Overall, this package of measures has resulted in almost a third of all new cars registered there in 2018 being electric vehicles. At this scale, however, the first side effects are also becoming apparent: being allowed to use bus lanes and thus drive past traffic jams used to be quite attractive. Today, the many electric vehicles together with the buses in their shared lane often jam up just as much as the combustion vehicles on neighboring lanes.
- Another example of targeted and long-term promotion of e-mobility is China. In order, first is to curb the dramatic increase in air pollution caused by the rapid growth of passenger cars in cities, second is to generate a new, internationally competitive industry sector, and third is to reduce dependence on oil imports from the Middle East; the twelfth 5-year plan issued in 2011 set the goal of the country becoming the global application and technology leader in the field of e-mobility within a few years. In order to stimulate domestic demand for electric vehicles, the purchase of private *new energy vehicles has* been subsidized since 2010, initially as part of a test phase in selected cities, with 60,000 yuan (about 6600 euros at the time) for BEVs and 50,000 yuan (about 5500 euros at the time) for PHEVs. Thanks in part to this subsidy, nationwide sales of New Energy Vehicles increased from about 8000 vehicles in 2011 to 1.3 million vehicles in 2018 (of which about 80 percent were BEVs), while total vehicle sales increased from 14.5 million to 23.7 million vehicles over the same period.
- There is no EU-wide agreement on the promotion of electric vehicles. Today, 24 of the member states have adopted national regulations in this regard, each of which consists of different proportions of a purchase subsidy, a waiver of vehicle tax, tax deductibility for company cars, and premiums for the scrapping of old vehicles. In Germany, for example, the purchase of a BEV or FCEV has been supported with a total of 4000 euros since 2016, while a PHEV is eligible for 3000 euros

here. In addition, both are exempt from vehicle tax for 10 years. Similar subsidy models with up to 5000 pounds (6618 euros) already exist since 2011 in the UK or since 2013 in Italy. From the French government, there are 6300 euros for buyers of BEV, PHEV, or FCEV and even 10,000 euros if a diesel vehicle at least 10 years old is scrapped at the same time. Here, too, electric vehicles are temporarily exempt from vehicle tax. The effect of the subsidy varies massively from country to country, with the share of electric vehicles in new registrations in 2018 ranging from almost 8 percent in Sweden to 2 percent in Germany.

- In the USA, the subsidy is made up of measures at federal and state level. At the state level, direct subsidies of varying amounts of up to 5000 US dollars (about 4500 euros) are paid out, as well as benefits for registration, the purchase and installation of a Wallbox, parking, and charging. In addition, there is permission to use the faster *"carpool lanes"*. The most attractive incentive package is offered in California; Alaska, Arkansas, Nebraska, Oklahoma, Maine, and South Carolina, on the other hand, do not offer any state incentives for electric vehicles as of today. At the federal level, a discount on income tax is currently granted for purchases depending on the size of the vehicle and the capacity of its battery, but since the change of government in 2017 at the latest, there has been no discernible political will to further promote e-mobility.

However, there is also criticism of the subsidy. For one thing, many of the electric vehicles on offer today are in the upper price segment. A purchase subsidy therefore always means promoting the acquisition of relatively expensive cars by correspondingly wealthy buyers. Secondly, a particularly attractive subsidy may also attract users of local public transport back to their own, now electrically powered, cars. Another point of criticism relates to the promotion of PHEVs. These can usually be driven in a pure hybrid mode, in which the electric motor is only used temporarily and the battery is not charged by plugging it into a charger but by the generator. Those who drive in this way therefore receive the subsidy without providing the intended reduction in pollutants and greenhouse gases in return. In China, the subsidy for some PHEV models has already been cancelled on the basis of a corresponding evaluation of vehicle data.

6.4.3.2   Promoting Alternatives to the Private Car

Compared to this diverse and extensive promotion of the purchase of BEVs, PHEVs, and FCEVs, the public incentivization of the use of mobility alternatives such as the use of public transport, private mobility services, cycling, or even walking is only discernible in a few, local approaches today – and this despite the fact that this clearly leaps further in terms of the long-term package of goals of reducing traffic volumes in cities, making road traffic safer, keeping the air clean, and curbing global warming.

As already stated in Sect. 5.4.6, Those who own a car and have thus already paid the purchase price or leasing fees feel little incentive to pay additional fees to use alternatives. In the meantime, many municipalities have understood that the only way to encourage car drivers to use public transport that is not based on bans or charges is to offer it not only cheaply but completely free of charge – and some European cities such as Monheim, Dunkirk, or Tallinn have already implemented this consistent form of promotion.

Financial incentives from the public sector for the use of private mobility services – whether taxis, ride-hailing, or car sharing – do not exist at all. Only the offer of the services is promoted here, and this only indirectly, for example, through their basic admission in the urban area or the reservation of parking spaces and charging stations. One reason for this reluctance may be that the municipalities want to prevent a migration of public transport users to private providers. As a consequence, the price level of private services still leads to a certain exclusivity. An example of a sensible use-related promotion would be the integration of private ride-sharing services into public transport at tariff level, which could then be used to close existing gaps in services in areas and time windows with low demand.

In addition to incentives aimed directly at users, employers in cities can also be encouraged and financially supported to make the use of alternatives to private cars as attractive as possible for their employees. These include, for example, a contribution to the costs of a monthly or annual ticket for public transport that can also be used privately, the granting of a mobility budget instead of the provision of a company car for executives, or also the promotion of the purchase of the so-called company bicycles, i.e., bicycles that are made available to employees at favorable conditions analogous to company cars.

In order to avoid the impending traffic collapse, municipalities are pursuing two different strategic thrusts: Making it more difficult to use one's own car through restrictive measures, and promoting alternatives to one's own car. Since the latter requires significantly more time and money, in the end the urgency and the available funds also decide which path is taken.

7

Looking Ahead

How Will Mobility Change and What Should We Do?

As has become clear in the previous chapters, the core question of this book, namely, how "we" will move in 2030 or later, is extraordinarily complex and cannot be answered by a single, concrete picture. Possible scenarios depend not only on assumptions regarding progress in the development of technical solutions but above all on societal trends and political framework conditions, as well as, in the final instance, on the financial resources available now and in the future in private households, companies, and public coffers – and thus turn out differently for each country and each conurbation. Only those who – whether out of ignorance, convenience, or calculation – ignore significant influencing factors and thus limit themselves to considering only small subareas will arrive at such simply structured prophecies as: "In the future, we will all only drive electric cars!" "There will be no more private cars in the cities!" "Electromobility will never catch on here!" Loudly voiced by politicians, activists, or individuals on the net and in public, such statements therefore serve less to provide orientation and strategic direction than consciously or unconsciously to polarize and provoke. Those who live and work in Berlin Mitte, for example, and no longer need their own car, may come to the conclusion and take the view that private cars are quite fundamentally unnecessary. By contrast, in the small towns in the

J. Weber, *Moving Times*, https://doi.org/10.1007/978-3-658-37733-5_7

Brandenburg hinterland, where you can't get to work or go shopping without your own car, the same view is probably regarded as ideological and unworldly even by the most innovation and sustainability oriented among the residents. However, despite all the differences and imponderables, some serious forecasts can be derived from the preceding consideration of the framework conditions.

7.1   There Is No Turning Back

Whether for economic reasons, convenience, or love of the familiar and the existing, many people ask themselves why everything can't just stay the way it is when it comes to mobility. The answer to this is relatively simple: not because new technical solutions make new alternatives possible but because a growing number of people worldwide are feeling more and more strongly and accepting less and less how the consequences of individual mobility in its current form are increasingly affecting their quality of life now and in the future.

First is because the emissions caused by mobility are becoming increasingly noticeable in society. Pollutant emissions and noise from combustion engines have always been in the public eye, but their effects on health and the environment are now even more concretely demonstrated and can be seen and experienced even more directly than before in the smog of Beijing and Delhi, for example. Above all, the greenhouse gases produced by the combustion of fossil fuels, such as carbon dioxide, have become much more prominent, at the latest since they are no longer just measured in the abstract and compared with limit values, but experienced first-hand – through the global climate phenomena that are increasingly occurring as a result of global warming.

On the other hand, and quite independently of the listed effects on the environment, because especially in metropolises and large cities, the increasing number of motor vehicles is leading more and more frequently to complete traffic gridlock while at the same time taking up more and more public space for roadways and parking spaces. Also, even if in Europe and the USA the number of deaths and injuries caused by traffic accidents is falling, in the booming cities of Asia, Africa, and South

America, it continues to rise with the increasing motorization of the population, where it is becoming another problem area of urban mobility that can be experienced very directly.

Both ultimately lead to the fact that the need for sustainable management and healthy living, which is constantly increasing in large parts of society anyway, is also increasingly encompassing mobility. More and more people are prepared to change their behavior in this sense and to accept restrictions to a certain extent. Parents are being "educated" by their children and older colleagues by younger ones. Surprisingly, even many people who do not want to or cannot change their own behavior themselves expect companies, authorities, and politicians to change accordingly.

7.2   Focus of the Change

So whether you like it or not, the attention that is so visibly being paid to environmental sustainability in particular today is not a short-term hype that will soon give way to the next. The noticeably worsening actual situation, the increase in scientifically proven findings, and the importance that the younger generation in particular attaches to the topic have heralded a sustainable social change. Together with the resulting laws and regulations as well as technical progress, this change is creating a significantly altered framework within which mobility solutions will lie in the future. The core question of mobility, how to get to one's destination as quickly, as comfortably and as cheaply as possible, must increasingly be expanded by the criterion of sustainability.

In the end, the individual decision as to how someone wants to reach their destination will continue to be influenced by the same factors as today, despite all the complexity of the topic: by availability, i.e., which options are available to choose from at any given time; by value for money, i.e., how fast, safe, and comfortable the options are in relation to the price called for them; and by how well the options correspond to the respective personal and social values. In each of these categories, specific developments can be identified that will influence the mobility of the future more than others.

7.2.1 Reducing Mobility Needs

Before even thinking about technical changes to vehicles and measures to reduce them, the absolute ideal way to get to grips with the problems of mobility is clearly to reduce the need for mobility itself.

A first approach to this lies in the decentralization of urban structures in metropolitan areas. How is it possible that quite a few residents of China Town or Little Italy in New York have not left their neighborhood for decades? Because they find everything they need to live there. Even if such ethnically oriented structures, which have grown over the years, are certainly not suitable as role models for other cities, where not only kindergartens and schools but also secondary schools, universities, and attractive jobs are available locally, only a few people have to commute by car or suburban train in the morning. Also, those who have not only the shops for everyday needs and general practitioners but also specialists and hospitals, restaurants, theaters, and cinemas in their neighborhood will also drive significantly less often into or across the city during the day and in the evening. This effect is supported by short-distance mobility offered in sharing, such as e-scooters or bicycles, which can be used to easily and conveniently expand the area within walking distance.

The planning and implementation of such decentralized structures is a mammoth task for the municipalities and can only be solved by them in the medium to long term. In the interests of a homogeneous social fabric, however, it must be avoided at all costs that certain age groups, ethnic groups, or social classes are concentrated in the city districts. In the end, decentralized structures not only relieve traffic but also distribute the attractiveness of a city more evenly from its centers to the surrounding areas, contrary to the typical historical structures – which not least also corresponds to the world of values of current social change.

A second approach to reducing the need for mobility is to strengthen the use of telecooperation across the board – for example, in the sense of teleworking, telelearning, or telemedicine. In many cases where information is primarily created or exchanged and no physical interaction is required, there is actually no need for the parties involved to meet in person. To listen to a lecture or process orders, a laptop or tablet is sufficient today; however, the prerequisite is that all parties involved embrace this change.

7.2.2 Consistent Continuation of Electromobility

Regardless of whether we are talking about cars, buses, or trains, society and legislators will continue to take appropriate measures, both vis-à-vis manufacturers and vis-à-vis operators and users, to ensure that vehicles on the roads become more environmentally friendly and also safer.

7.2.2.1 Distribution of Electric Vehicle Drives

BEVs are 100 percent emission-free in local use and can also be so in the well-to-wheel overall view – if operated with renewable energy. In addition, they use the primary energy input about twice as efficiently as vehicles with combustion engines. Purely electric vehicles have therefore already established themselves in urban traffic, not least because driving bans for vehicles with combustion engines are becoming increasingly foreseeable there. Electric vehicle drives will become established wherever daily driving distances are <500 kilometers, charging facilities are available, and a vehicle with a longer range can be used if required. An ideal application for BEVs is car-sharing or ride-hailing services.

The primary disadvantages and barriers to the purchase of electric vehicles are their comparatively short range, the need for charging stations, and the time required for charging. A great deal of effort is currently being put into reducing these disadvantages by car manufacturers, municipalities, and energy suppliers, so that further significant improvements can be expected here in the foreseeable future. The progress in development will then lead to more and more people no longer necessarily needing a conventional vehicle with an internal combustion engine but will also be able to cope with a then more powerful electric vehicle.

Plug-in hybrids are a good solution when longer ranges are often required, but it must or should also be possible to drive purely electrically in the inner city; they are therefore also a popular drive alternative today.

However, as the range of BEVs increases, the applications in which the additional combustion engine is still required will decrease. Since it will always be technically and economically costly to have two different drive systems in one vehicle, PHEVs are likely to become niche products in the long term.

Where greater range and generally higher energy requirements are needed, the direct generation of electrical energy from *on-board* stored hydrogen via a fuel cell represents a promising solution for energy storage. The range and refueling times of FCEVs available today are already in the range of 500–700 kilometers and 3–5 minutes, respectively, which are familiar from combustion vehicles. However, the breakthrough of FCEVs is hampered by two problems that have not yet been solved: On the one hand, the inherent fire and explosion risk of hydrogen creates high safety requirements for its storage in the vehicle; a technically simple and thus cost-effective but nevertheless safe solution does not yet exist. At the same time, this risk requires intensive product monitoring and support by the manufacturer. Unauthorized maintenance work or an unregulated used vehicle market is certainly unacceptable for FCEVs, and the technical behavior in the event of serious accidents is also not yet sufficiently safeguarded today. On the other hand, the development of an adequate infrastructure for hydrogen is still far more costly than that for charging electricity. Consequently, electric drives with fuel cells and hydrogen storage will therefore be found less in private cars but, apart from trucks in delivery traffic, primarily in buses for local and long-distance transport, which rely on the high range, can call at dedicated hydrogen filling stations on their route and are reliably maintained by the responsible operators.

The breakthrough of electric mobility will be additionally supported by the fact that electric motors are not only emission-free but also far more dynamic and comfortable than combustion engines. However, just how exciting electric driving can be cannot be gleaned from catalog data but must be experienced in person. The more people experience electric vehicles for themselves, the more this knowledge will spread.

7.2.3   Expansion of the Charging Infrastructure

The Achilles' heel of electromobility is and remains the availability of public charging facilities. The problem is not so much on the technical side. Powerful charging stations that can be used to charge electric vehicles quickly and yet without wear and tear are available on the market and are constantly being improved. Inductive charging will not play a role in a public infrastructure. Stationary inductive charging stations can be used as convenience solutions in the private sector, but dynamic inductive charging will not become established due to the extremely high implementation costs.

What is still missing today, however, is the right business model for the operators of charging stations. The crux here is that money cannot be earned from the sale of electricity on anywhere near the same scale as from the sale of petrol or diesel fuel. It is not for nothing that car manufacturers, energy suppliers, and local authorities have been at loggerheads for years over who should finance the necessary expansion of the charging infrastructure. Only when charging times reach the same magnitude as refueling times will it become interesting for filling stations to integrate the sale of electrical energy via charging stations into their business model.

There is widespread agreement today that electromobility will only really make sense ecologically if the electrical energy available at the charging station is also generated without emissions – and not, for example, through the irreversible combustion of fossil fuels such as oil, gas, or coal. There is far less agreement, however, on whether only renewable energy sources such as solar, wind, or hydroelectric power are emission-free or whether nuclear energy is also ecologically acceptable. The political attitude toward nuclear energy has become a kind of national question of faith among the industrialized nations: While Germany, Belgium, Spain, or Switzerland, for example, are already in the process of phasing out nuclear power, countries such as Great Britain, France, India, China, or the USA are building more nuclear power plants. Even in Japan, several reactors have now been put back into operation since the Fukushima disaster in 2011. The crucial point here is that investing in harnessing renewable energy only makes sense where nuclear power plants are not accepted as a clean and emission-free source of energy. The business model of building or operating solar, wind, and hydroelectric power plants, intermediate storage facilities, or transmission lines therefore stands or falls with a reliable political and public stance on nuclear energy.

7.2.4 Fewer Private Cars

Of course there will still be private cars in the future; there will still be no alternative for many people and in many situations in 2030, 2040, or 2050. However, especially – though not only – in the cities, in addition to traffic jams and a lack of parking spaces, rising charges for driving and parking, driving bans, and speed limits will mean that owning a car will become increasingly costly, subject to ever greater restrictions, less and less fun, and thus less and less desirable. The use of digital technologies will allow local authorities to automatically and thus completely and efficiently detect and sanction traffic violations or to stop them immediately and to issue precise instructions on which route to take and where to park. A right to drive faster than permitted, to use closed roads, to park in a no-parking zone – or even to freely choose the route to one's destination – is not guaranteed anywhere in the world and is becoming less and less acceptable to the majority.

However, the more adversity private vehicle ownership is subject to, the more people will abandon it and initially turn to car sharing or ride sharing services. This reduces the number of private vehicles, but the number of journeys and vehicles on the road remains. The switch from personal to third-party vehicles alone initially only reduces the need for parking spaces. It therefore only has a minor impact on overall emissions through the associated reduction in parking search traffic. In the long term, car sharing as well as ride hailing and taxis are therefore only the right alternative to owning a car in cases where a car is absolutely necessary.

A prerequisite for the acceptance of services such as car sharing or ride hailing is their guaranteed availability and attractiveness in terms of price. Both together will only be economically feasible in large cities in the foreseeable future, which is why an expansion of these services in small towns or in the countryside will only be possible to a very limited extent.

In any case, however, the following applies: In order to be seriously considered as an alternative to one's own car, a minimum availability of services in terms of space, time, and price is required. Public support not only for the users but also and above all for the operators of such services will help to reach this threshold more quickly. Also, only when such a basic supply has been established can differentiating offers be established on the market through special comfort or other service components.

People who are getting driven place completely different demands on vehicle features and functions such as comfort, dynamics, or connectivity than those who drive themselves. However, there are hardly any passenger cars today that are specifically designed for ride sharing services. If the boom in ride sharing services continues – and it can be assumed that it will – vehicle concepts designed for this purpose will represent a completely new and promising market segment for manufacturers.

Autonomous vehicles that drive entirely without a driver (Level 5) can be seen in this sense as the maximum level of passenger orientation. By eliminating the need for a driver, autonomous vehicle control makes ride hailing or ridesharing – at least in theory – not only cheaper but also safer for everyone involved in the traffic. Whether a driver will ultimately be appreciated and paid for as a convenience feature, however, remains to be seen when both systems are in real competition for passengers.

7.2.5 Fewer Cars on the Roads

The required reduction in traffic and environmental pollution is therefore quite obviously not achieved by using someone else's car instead of one's own but by reducing the total number of car journeys. Apart from the reduction in the need for mobility described above, which can only be achieved within narrow limits, this is only possible if more and more people use either a car together with others or other means of transport for their journeys.

The predominant approach here is to move people out of cars and into public transport, not only in cities but also for long-distance travel. The primary barrier to use for owners of a car that has already been paid for is the additional travel costs. In many cases, however, the service also needs to be made more attractive in terms of coverage of the route network, operating times, and safety of comfort. The latter includes, first and foremost, the adequate availability of room to stand or sit. The gaps in coverage in the service, for example, at night hours when demand is low or in the suburbs, can be closed flexibly and effectively by private ride-sharing services. Also, especially for these cases, the use of autonomous mancarrying drones is not as unrealistic as it may seem today.

Another approach is to share a car with several parties. Privately organized car pools are a particularly simple way here – although this presupposes that the journeys can be planned and that the start and destination are at least roughly the same. Such car pools are therefore particularly suitable for daily recurring journeys to the place of work or training and, in addition to sharing the operating costs, can become even more attractive through employer or municipal measures such as reserved parking spaces, lanes, or financial subsidies. Spontaneous car sharing, on the other hand, is only made possible by digital ride sharing services such as carpooling or ridesharing, in which suitable travel needs are combined according to time, route, and number of passengers. It is also true for these services that achieving the minimum availability of services required for a breakthrough can be effectively accelerated via public funding and, in some cases, can only be achieved with such support.

Depending on climatic and topographical conditions and travel distance, switching from cars to bicycles, e-bikes, or e-scooters is also an option. Appropriate infrastructural measures such as safe or covered cycle paths significantly increase acceptance here.

As far as the number of vehicles on the roads and at the roadside is concerned, more and more cities are pursuing not only the avoidance of a further increase but a noticeable reduction. Citizens who no longer own or use cars themselves also no longer see why so much public space should then continue to be used for driving and parking cars. The deconstruction of streets and parking spaces to cycle paths, green spaces, and public squares as a municipal measure makes car traffic even more burdensome and is perceived as a nuisance by the remaining car drivers. At the same time, it makes the city more attractive for pedestrians, cyclists, and public transport users – which further accelerates the social change toward a shift from cars to other modes of transport.

7.2.6 Growing Importance of Social Sustainability

Somewhat in the shadow of the surge in the importance of health and the environment, the importance of the various aspects of social sustainability is also increasing in large sections of society. Just as consumers now pay attention to fair trade in the case of food and to working conditions in the case of clothing, the conditions under which vehicles are manufactured or mobility services are provided are also coming under increasing critical public scrutiny in the mobility sector. The main focus here is on compliance with occupational safety and social standards vis-à-vis employees but also on compliance with commercial regulations such as anti-corruption.

As with environmental sustainability, vehicle manufacturers are also seen as responsible for social sustainability throughout the entire life cycle – i.e., for the value chain from the extraction of raw materials to vehicle assembly, for distribution and aftersales, and for all processes related to the disposal of the vehicle at the end of its life. Even in the provision of mobility services, the public is looking at compliance with social standards – as the public debate about the working conditions of Uber drivers shows, for example.

Negative reporting in terms of social sustainability has the potential to massively damage the image of brands and companies in a very short time. Whether it's toxic exposure of workers in cobalt mines, noncompliance with safety standards at suppliers, or drivers' working hours, today, social media brings questionable conditions to the public's attention in a matter of seconds, and this leads customers and sponsors if not to rejection then at least to critical questions.

Social and political pressure, however, are certainly not the only driving force here. More and more entrepreneurs see socially sustainable action as a personal duty and make it part of their corporate strategy. However, while it used to be considered noble to be modest about this, it will be out of place in the future, at least in the corporate environment. Regardless of the motivation behind the pursuit of social sustainability, transparency in this regard is becoming increasingly important for manufacturers and service providers in the mobility sector. Socially responsible action is of course the most important thing, but it must also be objectively assessable and verifiable – and therefore openly communicated.

7.3   Five Growth Areas: Where Is the Upward Trend?

Mobility today is therefore in the midst of change, in many places even dramatic change, in which some things that have long been right or cherished come to an abrupt end. At the same time, however, some things that have always been there are suddenly gaining significantly in importance, while new, unexpected, and surprising things are also emerging. At the end of this book, a brief look will be taken at five business areas which, in the author's opinion, will develop particularly positively over the next few years.

- Reduction of mobility needs:
  Cost reasons, environmental protection, time savings, and convenience, all of these together will lead to a further increase in demand for solutions that eliminate the need to travel and use vehicles. These can be cooperative technologies for teleworking, telelearning, telediagnosis, or virtual visits to the authorities but also the closing of gaps in the pedestrian offer within the neighborhood, be it shopping facilities for everyday necessities, specialist shops, or medical care.
- Expansion of local public transport:
  The demand is immense, but operation financed by the public sector rarely covers costs and in most cases is loss-making. However, anyone who can contribute with innovative vehicles or services to making public transport more available, more comfortable, safer, and thus more attractive and economical is a sought-after partner for a long time to come. Covering service gaps by integrating ride-hailing and especially ridesharing services but also cable cars or autonomous drones for passenger transport are just few examples.

- Premium ride sharing services:
  In parallel to public transport, the demand for private mobility services will also increase. In a first phase, a basic coverage must first be achieved that is necessary for acceptance as an alternative to the car and for which the number and price are in the foreground. Here, driverless, autonomous vehicles can help to achieve advantages in terms of availability and price. In a second phase of growth, there is then room for premium offerings that are placed above this basic offering and above public transport and are then much more economically attractive. Safety and comfort through passenger-oriented vehicles, first-class connectivity, and well-trained drivers are the main aspects of the attractiveness of such offers.
- Passenger-oriented vehicles:
  Vehicle concepts developed specifically for ride sharing services represent a new, secure growth area for automobile manufacturers, who are otherwise hard hit by the effects of change. These vehicle concepts can be used to present such services at significantly lower costs while at the same time offering significantly greater comfort and a broader range of services. The same applies to special accessories with which conventional vehicles can be temporarily or permanently optimized for use in ride-sharing services.
- Sustainability services:
  The increasing importance of ecological and social sustainability opens up an additional business field beyond vehicles and mobility services. On the one hand, there will be a massive increase in the need for advice on setting up but above all measuring, auditing, certifying, and communicating appropriate management systems; on the other hand, precisely these tasks to maintain the dual control principle cannot be performed by the manufacturers or service providers themselves.

Printed in the United States
by Baker & Taylor Publisher Services